Performance-Based
Fire Safety Design

Performance-Based Fire Safety Design

Morgan J. Hurley

Society of Fire Protection Engineers, USA

Eric R. Rosenbaum

Hughes Associates, USA

CRC Press
Taylor & Francis Group
Boca Raton London New York

CRC Press is an imprint of the
Taylor & Francis Group, an **informa** business

A SPON PRESS BOOK

Engineering A Fire Safe World

CRC Press
Taylor & Francis Group
6000 Broken Sound Parkway NW, Suite 300
Boca Raton, FL 33487-2742

First issued in paperback 2019

© 2015 by Taylor & Francis Group, LLC
CRC Press is an imprint of Taylor & Francis Group, an Informa business

No claim to original U.S. Government works

ISBN-13: 978-1-4822-4655-1 (hbk)
ISBN-13: 978-0-367-87027-0 (pbk)

Visit the Taylor & Francis Web site at
http://www.taylorandfrancis.com

and the CRC Press Web site at
http://www.crcpress.com

Contents

8 Smoke Control Design   131

9 Structural Fire Resistance   139

10 Fire Testing   149

Preface

This book was assembled from lecture notes that the authors created for courses on performance-based fire safety design at the University of Maryland, Worchester Polytechnic Institute, and California Polytechnic University during the last decade. The authors are indebted to constructive feedback (and corrections of errors) by countless students.

We created this book to serve two purposes: (1) as a textbook for academic programs on performance-based design, and (2) as a valuable reference for practitioners who wish to learn performance-based design or hone their skills. Some of the material that we provided in this book is a summary of information that can be found elsewhere, but other parts are new information that is an original contribution to the knowledge base.

We have been impressed by the accelerating rate at which performance-based fire protection is being accepted. As we note in Chapter 1, performance-based design is not new. What is new is the large number of authoritative references that are available to assist fire protection engineers and code officials alike. Many more are now willing to embrace performance-based design than was the case at the onset of our careers.

We hope that this book contributes in a meaningful way.

Acknowledgments

The authors gratefully acknowledge the review and constructive suggestions by Josh Dinaburg of Jensen Hughes.

About the Authors

Morgan Hurley is a project director with Aon Fire Protection Engineering and serves as adjunct faculty at the University of Maryland and California Polytechnic University. He holds bachelor's and master's degrees in fire protection engineering. Morgan chaired the NFPA Life Safety Code Technical Committee on Fundamentals, which developed the performance-based design option in NFPA 101 and NFPA 5000. Morgan is a recipient of the Committee Service Award for distinguished service in the development of NFPA codes and standards from the National Fire Protection Association. He is a licensed professional engineer and a fellow of the Society of Fire Protection Engineers.

Eric Rosenbaum is a registered fire protection engineer with over 20 years of experience in fire safety. Eric served as the chairman for the Society of Fire Protection Engineers (SFPE) Task Group on Performance-Based Analysis and Design. He is also a fellow of the SFPE and a registered professional engineer. Eric is an adjunct professor at the University of Maryland. As a member of the National Fire Protection Association (NFPA) and its board of directors, Eric chaired the Life Safety Code Technical Committee on Fire Protection Features for 9 years. He is on the NFPA Technical Committee on Health Care Occupancies and Technical Correlating Committee of the NFPA Life Safety Code, and is also a member of the Technical Committee for Board and Care Facilities (Assisted Living Facilities) and Means of Egress of the NFPA Life Safety Code. In 2011, Eric received the Committee Service Award for distinguished service in the development of NFPA codes and standards from the National Fire Protection Association, and has been the recipient of the Hat's Off Award from the Society of Fire Protection Engineers multiple years.

Chapter 1

Introduction

WHAT IS PERFORMANCE-BASED DESIGN?

The *SFPE Engineering Guide to Performance-Based Fire Protection* (SFPE, 2007) defines *performance-based design* as "an engineering approach to fire protection design based on (1) agreed upon fire safety goals and objectives, (2) deterministic and/or probabilistic analysis of fire scenarios, and (3) quantitative assessment of design alternatives against the fire safety goals and objectives using accepted engineering tools, methodologies, and performance criteria."

This definition identifies three key attributes of performance-based design. The first is a description of the desired level of fire safety in a building (or other structure) in the event of a fire. The second attribute includes definition of the design basis of the building. The design basis is an identification of the types of fires, occupant characteristics, and building characteristics for which the fire safety systems in the building are intended to provide protection. In the vernacular of performance-based design, these fires are referred to as design fire scenarios. The third element involves an engineering analysis of proposed design strategies to determine whether or not they provide the intended level of safety in the event of the design fire scenarios.

Nelson (1996) identifies four types of performance:

Component performance. Component performance identifies the intended performance in fire of individual building systems or components, such as doors, structural framing, or individual protection systems such as detection. In component performance analysis, individual components and systems are designed in isolation, without considering how their performance may impact, or be impacted by, the performance of other systems or components. Any system or component that meets the stated performance would be considered to be acceptable.

An example of a component performance-based approach would be a structural element that is designed to achieve a 1 h rating when exposed to the standard fire. In this case, the intended performance would involve maximum acceptable point and average temperatures, and the design fire scenario would be the ASTM E-119 time-temperature curve. While building codes typically require this performance to be achieved through fire testing, calculation methods are available as well (ASCE, 2005). Any assembly that achieves the intended performance when exposed to the design fire scenario would be considered acceptable.

Another example would be an individual sprinkler used in a sprinkler system. Sprinkler design standards and approval standards might require a maximum actuation temperature and thermal response characteristics. Any sprinkler that met the performance identified would be acceptable.

It is noteworthy that the codes and standards that govern fire-resistant structural elements and sprinklers contain specific requirements that are not performance based, such as limitations on the types of materials that can be used in fire-resistant assemblies and sprinklers.

Environment performance. Environment performance involves identification of the maximum permissible fire conditions within a building or portion thereof. The specification of environmental conditions could involve temperature, heat flux, or products of combustion. Environmental performance approaches identify conditions that are tolerable if a fire were to occur. It is not possible to include fire prevention strategies within an environmental performance approach.

An example of an environmental performance approach would be a requirement that the smoke layer within an atrium not descend below a given elevation above the highest occupied level. Any design that could achieve this criterion would be acceptable, and the performance requirement does not specify or limit how this can be achieved.

Threat potential performance. Threat potential performance involves identification of the maximum acceptable threat to life, property, business continuity, or the natural environment. Unlike environmental performance requirements, which involve statements of maximum acceptable conditions in the environment surrounding items that are desired to be protected from fire, threat potential performance involves a statement of the maximum tolerable conditions of the item or items being protected.

An example of a threat potential performance requirement would be a fractional effective incapacitation dose (see Chapter 6). Another example would be an identification of the maximum permissible temperature of an object. As with environmental performance, threat potential performance identifies conditions that are tolerable if a fire were to occur.

Risk potential performance. In risk potential performance, the summation of the products of probabilities of occurrence of fire events and their consequences are specified. An example of a risk potential performance requirement would be that the average permissible property loss in a facility resulting from fire must not exceed an average of $10,000 in value per year. When applying this type of approach, a designer would evaluate all possible fire events and their potential consequences. This can be expressed mathematically as (SFPE, 2007)

$$Risk = \sum Risk_i = \sum (Loss_i \bullet P_i)$$

where $Risk_i$ is risk associated with scenario i, $Loss_i$ is loss associated with scenario i, and P_i is probability of scenario i occurring.

Nelson (1996) also identifies the typical prescriptive approach, which he defines as specification. Specification involves strict definition of dimensions, construction methods, and other features. An example of specification would be some of the requirements in the *Life Safety Code* (NFPA, 2012a) applicable to stairway construction. The *Life Safety Code* identifies specific dimensional requirements for stairs and handrails.

HISTORY OF PERFORMANCE-BASED DESIGN OF BUILDINGS

As can be seen in the preceding discussion, performance-based requirements are in no way new. Early (pre-1900s) fire protection requirements largely fit into the category of specification, with such requirements including the permissible materials from which building exteriors could be constructed or the minimum acceptable spacing between buildings. However, most modern building and fire code requirements have some element of performance associated with them.

Performance-based approaches for designing building fire protection can be traced to the early 1970s, when the goal-oriented approach to building fire safety was developed by the U.S. General Services Administration (Custer and Meacham, 1997). Other major developments in performance-based design include the following:

- Publication of the performance-based British Regulations in 1985
- Publication in 1988 of the first edition of the *SFPE Handbook of Fire Protection Engineering*
- Publication of the performance-based New Zealand Building Code in 1992 and the New Zealand *Fire Engineering Design Guide* in 1994
- Publication of the Performance Building Code of Australia and the Australian *Fire Engineering Guidelines* in 1995

- Publication of the *Performance Requirements for Fire Safety and Technical Guide for Verification by Calculation* by the Nordic Committee on Building Regulations in 1995
- Publication of the performance option in the NFPA *Life Safety Code* in 2000
- Publication of the *SFPE Engineering Guide to Performance-Based Fire Protection Analysis and Design of Buildings* in 2000
- Publication of the Japanese performance-based Building Standard Law in 2000
- Publication of the *ICC Performance Code for Buildings and Facilities* in 2001
- Publication of the performance option in the NFPA *Building Code* in 2003

The documents only represent the formalization of performance-based design. Performance-based design has long been practiced through the use of equivalency or alternate methods and materials clauses found in most, if not all, prescriptive codes and standards. These clauses permit the use of approaches or materials not specifically recognized in the code, provided that the approach or material can be demonstrated as providing at least an equivalent level of safety as that required by the code or standard.

However, equivalency or alternate methods and materials clauses typically do not provide any detail as to how an equivalent level of safety can be achieved. Therefore, the approaches used by individual designers or regulatory officials were frequently developed on an ad hoc basis, with approaches varying among designers and regulatory officials. The effect of the documents identified in the preceding text was to standardize the practice of performance-based design.

The evolution of performance-based design has followed an evolution in the quantitative understanding of fire. Before fire science was well understood, proven technologies would be codified into regulations. Similarly, as major fires occurred, and the causes and contributing factors of those fires were identified, codes and standards were modified to prevent similar major fires from occurring in the future.

Specification codes have two disadvantages:

- They can only protect against events of a type that have occurred in the past. Major fires are low-probability, high-consequence events. Because of their stochastic nature, some types of rare events have not yet occurred.
- They can stifle innovation. By specifying certain types of methods and materials, it can be difficult to introduce new methods and materials into the marketplace.

As the science of fire became better understood, performance-based fire protection design has become possible. Other engineering disciplines have evolved in a similar manner—as the underlying science became better understood, their design approaches became more performance based.

ADVANTAGES AND DISADVANTAGES OF PERFORMANCE-BASED DESIGN OF BUILDINGS

Performance-based design offers a number of advantages and disadvantages over specification-based prescriptive design. As the design approach used moves from specification based toward risk based (see the "What Is Performance-Based Design" section above), these advantages and disadvantages are magnified.

Advantages:
- Performance-based design allows the designer to address the unique features and uses of a building. An example would be the types of stores that can be found in a shopping mall. Each might have an identical occupancy classification under prescriptive building and fire codes, and hence require similar fire protection strategies. However, the stores could contain significantly different fire hazards. Some could contain flammable liquids, while others might contain few or no combustible items at all. A corollary to this advantage is increased cost-effectiveness of performance-based designs.
- Performance-based design promotes a better understanding of how a building would perform in the event of a fire. Compliance with prescriptive codes and standards is intended to result in a building that is "safe" from fire. However, what constitutes safe is generally not defined. Similarly, the types of fires against which the building is intended to achieve fire safety are not identified. While most common fire scenarios would likely result in acceptable performance, the low-frequency scenarios that are not envisioned may not.

   Two fire scenarios can be used to illustrate this. Carelessly discarded smoking materials would likely be within the design basis for a code that is intended to apply to a high-rise residential building. However, a gasoline tank truck that accidentally crashes into the building's lobby likely is not. Within these two extremes are a large range of possible events. A corollary to this advantage is that increased thought and engineering rigor is brought to solving fire protection problems.

Disadvantages:

- Performance-based design requires more expertise to apply and review than does prescriptive-based design. Application of prescriptive codes only requires the selection of building features and systems that meet the code's requirements. Verification of the acceptability of a prescriptive-based design is equally straightforward. Performance-based design can take more time to conduct and review than prescriptive-based design.

- Performance-based design can be more sensitive to change than prescriptive-based design. Changes in use of a building or portion thereof can result in unacceptable performance in the event of a fire if the effect of the change on fire safety is not contemplated in the design. With prescriptive-based designs, changes in use may be acceptable if the portion modified stays within the original occupancy classification. This is not to say that prescriptive designs are completely tolerant to changes; even if a modification remains within the original occupancy classification, some types of changes could result in the modification not being compliant with prescriptive codes. For example, movement of walls during tenant renovations in an office building could result in the sprinkler system no longer being in compliance with governing codes and standards. If a building is designed according to a performance basis, then some changes in use may result in increased vulnerability in the event of a fire.

The process that is identified in the subsequent section provides methods of overcoming the limitations.

PROCESS OF PERFORMANCE-BASED DESIGN

The *SFPE Engineering Guide to Performance-Based Fire Protection* (2007) provides a process, or framework, for performance-based design. This process is identified in the flowchart in Figure 1.1. The process is intended to be flexible, so that it can be tailored to the individual requirements of individual performance-based design projects.

This process identifies the steps that are involved in performance-based design, without specifying which methods or models should be used in the development or evaluation of a specific design. While widely used references are identified within the guide, the references are not intended to be endorsed or comprehensive.

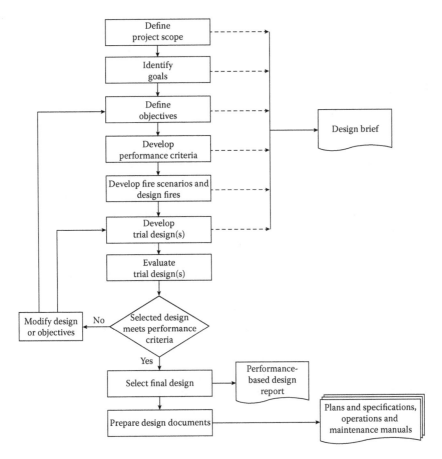

Figure 1.1 Performance-based design process. (Used with permission of Society of Fire Protection Engineers, copyright © 2007.)

Defining the Project Scope

The performance-based design process identified in the *SFPE Engineering Guide to Performance-Based Fire Protection* begins with developing the project scope. Project scopes for performance-based designs are not significantly different from project scopes for prescriptive-based designs, although unique features may be identified that might be difficult or impossible to achieve through strict compliance with prescriptive-based codes.

The project scope identifies the portions of a building or facility that will be considered by the design, the desired features of the design, the intended characteristics of the building, and the regulations that are applicable. The scope also includes identification of the project stakeholders—those that have an interest in the success of the design. Stakeholders may include building owners or their representatives, regulatory authorities, insurance providers, building tenants, fire officials, or other parties. From the scope, a clear understanding can be gained of the needs of the project.

Identifying Goals

Once the scope is identified, the next steps involve the definition of goals and objectives for the design project. The *SFPE Engineering Guide to Performance-Based Fire Protection* (2007) defines *goals* as the "desired overall fire safety outcome expressed in qualitative terms." Goals are intended to be stated in broad terms that can easily be understood by people who may not have engineering expertise. The purpose of identifying goals is to facilitate understanding and agreement on how the building is intended to perform in the event of a fire.

Laypeople would likely not be able to understand the significance of keeping the upper-layer temperature below a certain threshold, but they could understand what it means to provide for life safety in the event of a fire.

The *SFPE Engineering Guide to Performance-Based Fire Protection* identifies four fundamental goals for fire safety: life safety, property protection, mission continuity, and environmental protection. While these types of statements are entirely qualitative in nature, they point to the direction of the design process. For example, an unattended, fully automated warehouse may have property protection and mission continuity as its primary design goals. A hotel would likely have life safety as its primary fire safety goal.

Goals can come from a variety of sources. Some codes identify goals. For example, NFPA 101, the *Life Safety Code*, specifies the following goals (NFPA, 2012a):

Fire and Similar Emergency

The goal of this *Code* is to provide an environment for the occupants that is reasonably safe from fire and similar emergencies by the following means:

(1) Protection of occupants not intimate with the initial fire development
(2) Improvement of the survivability of occupants intimate with the initial fire development

Crowd Movement

An additional goal is to provide for reasonably safe emergency crowd movement and, where required, reasonably safe nonemergency crowd movement.

NFPA 5000, the *Building Construction and Safety Code* (NFPA, 2012c), provides the following goals:

Goals

The primary goals of this *Code* are safety, health, building usability, and public welfare, including property protection as it relates to the primary goals.

NFPA 5000 specifies more goals than NFPA 101 does, which is due to the fact that NFPA 5000 has a broader scope than NFPA 101. NFPA 101 only addresses life safety, while NFPA 5000 addresses many additional aspects of building safety. The *ICC Performance Code for Buildings and Facilities* (ICC, 2012) identifies goals that are similar to those contained in NFPA 5000.

Designs that comply with the prescriptive option of NFPA 101 or NFPA 5000 are deemed to comply with the goals specified by those codes. Similarly, designs that comply with the International Code Council (ICC) family of codes are deemed to comply with the goals of the *ICC Performance Code for Buildings and Facilities*. However, designers that prepare performance-based designs would have to demonstrate that the designs achieve the goals of the applicable code.

In some cases, project stakeholders may specify their own goals. See the "Application of Performance-Based Design" section.

As qualitative statements, goals are insufficient to judge the adequacy of a design. Therefore, they will have to be quantified as measurable values. The next two steps of the process outlined in the *SFPE Engineering Guide to Performance-Based Fire Protection* are intended to facilitate translating these broad statements into specific numerical criteria that can be predicted using engineering methods.

Defining Objectives

The next step in this process is the development of objectives. The *SFPE Engineering Guide to Performance-Based Fire Protection* identifies two types of objectives: stakeholder objectives and design objectives. Stakeholder objectives provide greater detail of the tolerable levels of damage the than goals do. Stakeholder objectives might be expressed in terms of maximum allowable levels of injury, damage to property, damage to critical equipment, or length of loss of operations.

After the stakeholder objectives have been developed and agreed upon, the *SFPE Engineering Guide to Performance-Based Fire Protection* recommends developing design objectives. Design objectives focus on the items that are intended to be protected from fire, and describe the maximum or minimum acceptable conditions necessary to achieve the stakeholder objectives.

As with goals, stakeholder objectives could be specified by a performance-based code. For example, NFPA 101 (2012a) provides the following objectives:

Occupant Protection

A structure shall be designed, constructed, and maintained to protect occupants who are not intimate with the initial fire development for the time needed to evacuate, relocate, or defend in place.

Structural Integrity

Structural integrity shall be maintained for the time needed to evacuate, relocate, or defend in place occupants who are not intimate with the initial fire development.

Systems Effectiveness

Systems utilized to achieve the goals shall be effective in mitigating the hazard or condition for which they are being used, shall be reliable, shall be maintained to the level at which they were designed to operate, and shall remain operational.

NFPA 5000 provides additional objectives resulting from the additional goals of the code.

If they are not specified by a code, stakeholder objectives will need to be developed by the engineer in consultation with project stakeholders based on the goals. In most cases, design objectives would be developed by an engineer based on the goals and stakeholder objectives agreed to by the stakeholders.

Developing Performance Criteria

Performance criteria are threshold values that, if exceeded, indicate that unacceptable damage has occurred. While design objectives provide more detail than the goals or stakeholder objectives, they are not sufficiently detailed for the evaluation of trial designs.

Performance criteria might include temperatures of materials, gas temperatures, smoke concentration or obscuration levels, carboxyhemoglobin concentrations, or radiant heat flux levels. Performance criteria should be predictable with engineering tools such as fire models.

The *SFPE Engineering Guide to Performance-Based Fire Protection* divides the types of performance criteria that may need to be developed into two categories: life safety criteria and non-life safety criteria.

Life safety criteria address the survivability of people exposed to fire or fire products. The values selected as performance criteria might vary depending upon the physical and mental conditions of building occupants and length of exposure. Performance criteria may need to be developed in the areas of thermal effects to people (e.g., exposure to high gas temperatures or thermal radiation), toxicity of fire products, or visibility through smoke.

Non-life safety criteria may need to be developed to assess the achievement of goals relative to property protection, mission continuity, or environmental protection. Performance criteria relative to these goals may relate to thermal effects, such as ignition, melting, or charring, fire spread, smoke damage, fire boundary damage, structural integrity, damage to exposed items, or damage to the environment.

Given that performance criteria can vary widely depending upon the specific design situation, the *SFPE Engineering Guide to Performance-Based Fire Protection* does not provide specific performance criteria. Rather, the guide identifies a number of reference sources that can be used to assist with the development of design-specific performance criteria.

Table 1.1 contains examples of goals, objectives, and performance criteria.

Table 1.1 Examples of Stakeholder Objectives, Design Objectives, and Performance Criteria

Fire Protection Goal	Stakeholder Objective	Design Objective	Performance Criteria
Minimize fire-related injuries and prevent undue loss of life	No loss of life outside of the room or compartment of fire origin	Prevent flashover in the room of fire origin	Upper layer temperature not greater than 200°C
Minimize fire-related damage to the building and its contents	No significant thermal damage outside of the room or compartment of fire origin	Minimize the likelihood of fire spread beyond the room of fire origin	Upper layer temperature not greater than 200°C
Minimize undue loss of operations and business-related revenue due to fire-related damage	No downtime exceeding 8 h	Limit the smoke exposure to less than would result in unacceptable damage to the target	HCl not greater than 5 ppm; particulate not greater than 0.5 g/m³
Limit environmental impacts of fire and fire protection measures	No water contamination by fire suppression water runoff	Provide a suitable means for capturing fire protection water runoff	Impoundment capacity at least 1.20 times the design discharge

Source: SFPE, 2007.

Some performance-based codes provide performance criteria. NFPA 101 (2012a) provides the following performance criterion:

Performance Criterion

Any occupant who is not intimate with ignition shall not be exposed to instantaneous or cumulative untenable conditions.

Since "instantaneous or cumulative untenable conditions" is not defined, this performance criterion is more akin to an objective. However, additional specificity can be found in the annex of NFPA 101. The options outlined in the annex deal with prevention of incapacitation from smoke or prevention of exposure to smoke.

In many cases, the engineer will need to develop performance criteria from the goals and objectives. To develop performance criteria, it is necessary to understand the mechanism of harm to the object being protected. Chapter 6 addresses the mechanisms of harm to people in detail.

Developing Fire Scenarios

There are two sets of information that are needed to evaluate whether a trial design is acceptable. One is the performance criteria. The second is the design fire scenarios, which describe the types of fires for which a design is intended to provide protection.

The *SFPE Engineering Guide to Performance-Based Fire Protection* suggests a two-step process for identifying design fire scenarios. The first step is considering all possible fire scenarios that could occur in the building or portion of the building that is within the scope of the design. The second step is to reduce the population of possible fire scenarios into a manageable set of design fire scenarios.

Both fire scenarios and design fire scenarios comprise three sets of characteristics: building characteristics, occupant characteristics, and fire characteristics. Building characteristics describe the physical features, contents, and ambient environment within the building. They can affect the evacuation of occupants, growth and spread of fire, and movement of combustion products. Occupant characteristics determine the ability of building occupants to respond and evacuate during a fire emergency. Fire characteristics describe the history of a fire scenario, including first item ignited, fire growth, flashover, full development, and decay and extinction.

The *SFPE Engineering Guide to Performance-Based Fire Protection* identifies a number of methods that can be used to identify possible scenarios. These include:

- Failure modes and effects analysis, where the different types of failures that could occur are studied, and the effects of those failures are analyzed.
- Failure analysis, where potential causes of failures are identified and the expected system performance is investigated.
- "What if" analysis, where expert opinion is used to consider possible events and the consequence of those events.
- Historical data, manuals, and checklists, where past events in a building or a similar building are studied to consider whether similar events could occur in the building that is being designed or modified. Manuals and checklists can be studied to consider warnings, cautions, or operational sequences that could lead to a fire if not followed.
- Statistical data of fires across broad classifications of buildings.
- Other analysis methods, such as hazards and operability studies, preliminary hazard analysis, fault tree analysis, event tree analysis, cause-consequence analysis, and reliability analysis.

Given the large number of possible fire scenarios for a given performance-based design project, it is usually necessary to reduce the possible fire scenario population to a manageable number of design fire scenarios for evaluating trial designs. For most design projects, this can be accomplished in part by excluding scenarios that are highly unlikely to occur or that would result in an acceptable outcome regardless of the trial design strategy that is used. However, for a fire scenario to be excluded from further analysis because it is considered too unlikely, all stakeholders must recognize and accept that if the scenario were to occur, an unacceptable outcome may result.

Another method of reducing the number of fire scenarios is to select bounding scenarios, where if the performance criteria can be achieved in these scenarios, it can be safely assumed that they would be achieved in the scenarios that are not specifically considered.

For risk-based analyses, it would only be acceptable to exclude a fire scenario from further consideration if it could be established that no design could handle the scenario. Methods of reducing the number of scenarios that must be analyzed in risk analyses will be discussed in more detail in Chapter 2.

Some performance-based codes provide fire scenarios. Even when such a code is applicable to a design, the fire protection engineer should work with project stakeholders to determine if there are other scenarios that should be considered.

Fire scenarios will be discussed in greater detail in the chapters on design fires.

Development of Trial Designs

Trial designs are fire protection strategies that are intended to achieve the goals of the project. To be considered acceptable, trial designs must achieve each of the performance criteria when subjected to the design fire scenarios.

The *SFPE Engineering Guide to Performance-Based Fire Protection* groups the types of methods that might be used in trial designs into six subsystems. Attributes from one or more subsystems would be used in a trial design. The six subsystems identified in the guide are:

- Fire initiation and development, where methods are used to reduce the likelihood that ignition would occur or reduce the rate of fire development if a fire were to occur
- Spread, control, and management of smoke, where smoke hazards are reduced by limiting smoke production, controlling smoke movement, or reducing the amount of smoke after it has been produced
- Fire detection and notification, where the presence of a fire would be detected for purposes of notifying building occupants or fire responders, or to activate a fire suppression system
- Fire suppression, including automatic or manual systems
- Occupant behavior and egress, where the travel to a place of safety prior to the onset of untenable conditions is facilitated
- Passive fire protection, including limiting fire spread through construction or preventing premature collapse of all or part of a structure

When developing trial designs, the engineer should refer back to the goals of the analysis and decide on what types of strategies would best achieve those goals. The NFPA *Fire Safety Concepts Tree* (NFPA, 2012b) can assist with the development of trial design strategies. The top branches of the tree may closely align with the objectives of the design. In these cases, the protection methods that are identified below the objectives that align with the design goals could be used as trial designs.

When using the *Fire Safety Concepts Tree*, users need to consider *and* gates and *or* gates. Where there is an *and* gate, each of the elements on the subordinate branches must be satisfied. Conversely, only one of the elements below an *or* gate must be achieved to satisfy the superior requirement.

Trial design strategies involve the same types of building components and systems that would be included in prescriptive designs. In fact, compliant prescriptive system designs may be appropriate as part of a trial design strategy. However, in some cases, augmented performance may be needed beyond that which would be achieved by a prescriptive-compliant system.

Fire Protection Engineering Design Brief

The preceding steps constitute the qualitative portion of the design, and agreement of all stakeholders should be attained prior to proceeding to the quantitative analysis. A mechanism that is suggested by the *SFPE Engineering Guide to Performance-Based Fire Protection* for achieving this agreement is a fire protection engineering design brief.

Evaluating and formally documenting performance-based designs can require extensive effort, and if fundamental aspects of the design change after detailed evaluation, significant rework may be required. For example, if a design is completed and evaluated based on achieving life safety goals, and after the design is evaluated property protection goals are identified, then effort previously expended may be wasted. Similarly, if project stakeholders insist on certain types of design strategies being used, then these should be identified before other types of design strategies are developed and evaluated. The purpose of the fire protection engineering design brief is to facilitate agreement on the qualitative portions of the design prior to conducting detailed engineering analysis.

The contents of the fire protection engineering design brief will typically include the project scope, goals, objectives and performance criteria, design fire scenarios, and trial design strategies proposed for consideration. The form of the fire protection engineering design brief is intended to be flexible, based on the needs of the project and the relationship of the engineer performing the design to other stakeholders. In some cases, a verbal agreement may be sufficient. In other cases, formal documentation, such as minutes of a meeting or a document that is submitted for formal review and approval, may be prudent.

Once the design team and stakeholders have agreed on the approach that is proposed for the performance-based design, the detailed analysis work begins. This includes quantification of the design fire scenarios, evaluation of trial designs, and development of project documentation.

Quantification of Design Fire Scenarios

After the design fire scenarios have been agreed upon by the stakeholders, they need to be quantified to the extent necessary to evaluate the trial designs. The building characteristics, occupant characteristics, and fire characteristics will each need to be quantified as needed to provide the details necessary to evaluate the trial designs.

While several types of characteristics are identified in the *SFPE Engineering Guide to Performance-Based Fire Protection*, they are intended to include all of the types of characteristics that might need to be quantified. However, for most design situations it will not be necessary to quantify all of the characteristics.

Quantification of Building Characteristics

Building characteristics describe the physical features, contents, and internal and external environments of the building. Building characteristics can affect the evacuation of occupants, the growth and spread of fire, and the movement of combustion products. The *SFPE Engineering Guide to Performance-Based Fire Protection* identifies the following building characteristics that may need to be quantified:

- Architectural features, such as compartment geometry, interior finish, construction materials, and openings
- Structural components, including any protection characteristics
- Fire load
- Egress components
- Fire protection systems
- Building services, such as ventilation equipment
- Building operational characteristics
- Firefighting response characteristics
- Environmental factors (interior and exterior temperatures, wind speeds, etc.)

Occupant Characteristics

For any design in which life safety or occupant response is considered, it will be necessary to consider the occupant characteristics. The *SFPE Engineering Guide to Human Behavior in Fire* (SFPE, 2003) identifies the following fundamental occupant characteristics that could influence the response of building occupants to a fire:

- Population (number and density)
- Alone or with others
- Familiarity with the building
- Distribution and activities
- Alertness
- Physical and cognitive ability
- Social affiliation
- Role and responsibility
- Location
- Commitment
- Focal point
- Occupant condition
- Gender
- Culture
- Age

Occupant characteristics provide information as to how people might respond when subjected to fire cues, where fire cues include seeing fire or smoke, smelling fire, hearing a fire alarm audible signal, or other cues. Occupant characteristics will be discussed in further detail in Chapter 6.

Design Fire Curves

Fire characteristics will typically be quantified as design fire curves, which provide a history of the size of a fire as a function of time. Typically, the size of a fire is measured in terms of heat release rate. Figure 1.2 shows an example of a design fire curve.

The *SFPE Engineering Guide to Performance-Based Fire Protection* divides design fire curves into five stages. Depending on the scope of the design, it may not be necessary to quantify each stage of the design fire curve. For example, it may only be necessary to quantify the growth stage for evaluation of a detection system. Similarly, evaluation of structural integrity may only require quantification of the fully developed stage. The guide provides suggestions on how to quantify each stage of the design fire curve as follows:

> For most designs, ignition will be assumed to occur. Typically, the design team will consider different first items ignited. If information is known about an item and an energy source, it is possible to predict whether the item will ignite.
>
> After an item ignites, the fire may grow in size. The rate at which a fire grows is a function of the first item ignited and the location of the item within a compartment. As the fire grows, additional items may be ignited and the fire may spread outside of an enclosure.

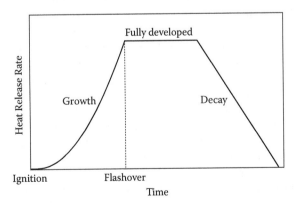

Figure 1.2 Sample design fire curve. (Used with permission of Society of Fire Protection Engineers, copyright © 2007.)

Flashover occurs when all combustible items within an enclosure ignite. Compartment geometry, compartment ventilation, fire heat release rate, and the thermal properties of the enclosure influence whether and when flashover occurs in a compartment.

If there is no intervention, a fire may reach a maximum size, which is a function of either the amount of fuel in the compartment or the amount of available ventilation. The fully developed stage of the fire is typically used to determine the effect of radiation through openings, failure of the structure, fire spread to other enclosures or failure of compartmentation. Fires will decay and eventually burn out. Decay can occur due to depletion of fuel, lack of ventilation, or manual or automatic or manual suppression.

Quantification of design fire scenarios will be discussed in greater detail in Chapter 4.

Evaluating Trial Designs

Evaluation is the process of determining if a trial design meets all of the performance criteria in each of the design fire scenarios. The *SFPE Engineering Guide to Performance-Based Fire Protection* states that the level, or detail, of an evaluation is a function of factors such as the complexity of geometry, level of subsystem interaction, and margin between evaluation output and the performance criteria. In some cases, a relatively simple evaluation may be appropriate, while in others, an in-depth evaluation would be required.

The following levels of evaluation are identified in the guide:

- **Subsystem.** A subsystem performance evaluation typically consists of a simple comparative analysis in which the performance of a design that involves modification of a single component or subsystem (e.g., egress, detection, suppression, fire resistance, etc.) is compared to the performance of a similar component or subsystem. This type of analysis is frequently employed when using the equivalency provision in a prescriptive code. For an alternate design strategy to be acceptable, it must provide equal or greater performance than that which is required by the code or standard.
- **System.** A system performance evaluation might consist of a comparison to prescriptive requirements or an analysis based on specific performance requirements. A system performance evaluation is used when more than one fire protection system or feature is involved. It is more complex than a subsystem evaluation because the analysis must account for the interaction between various subsystems.
- **Whole building.** In a building performance analysis, all subsystems used in the protection strategy and their interactions are considered. A performance-based design that analyzes total building fire safety

can provide more comprehensive solutions than subsystem or system performance analysis because the entire building-fire-target (where targets are the items being protected, such as people, property, etc.) interaction is evaluated.

The levels of performance described above describe the complexity of a design, whereas the types of performance identified by Nelson describe approaches that a code or standard could use to state desired fire performance.

Typically, engineering tools such as fire models will be used to evaluate trial designs. The tool(s) that are selected must provide information that can be used to determine whether or not the performance criteria have been achieved.

Additionally, as noted in the guide, uncertainty is always present in any design or analysis. Methods of dealing with uncertainty will be presented in Chapter 12.

Documentation

Following completion of the evaluation and selection of the final design, thorough documentation of the design process should be prepared. This documentation serves three primary purposes: (1) to present the design and underlying analysis such that it can be reviewed and understood by project stakeholders, such as regulatory officials, (2) to communicate the design to the tradespeople who will implement it, and (3) to serve as a record of the design in event that it is modified in the future or if forensic analysis is required following a fire.

The *SFPE Engineering Guide to Performance-Based Fire Protection* provides detailed descriptions of the types of documentation that should be prepared by the design team. This includes the documentation associated with the fire protection engineering design brief (discussed previously), a performance-based design report, specifications and drawings, and operations and maintenance manuals.

The guide suggests that a detailed performance-based design report should be prepared that describes the quantitative portions of the design and evaluation. Every model or calculation method that was used should be documented, including the basis for selection of the model or calculation method. Similarly, any input data for the model or calculation method should be documented, including the source of the input data and the rationale of why the data are appropriate for the situation being modeled.

All fire protection analyses have some uncertainty associated with them (as will be discussed in Chapter 12). This uncertainty may come from limited ranges of applicability of or simplifications within models or calculation methods, applicability of data sources to the scenarios modeled, limitations of scientific understanding, or other sources. The design should include methods of compensating for this uncertainty, and how this was accomplished should be documented.

As with prescriptive designs, performance-based designs use specifications and drawings to communicate to tradespeople how to implement the design. However, master specifications may not be applicable to performance-based designs without significant editing. Similarly, any features of a design that differ from typical prescriptive designs should be clearly identified on drawings.

One feature of documentation of performance-based designs that differs significantly from prescriptive-based designs is the operations and maintenance manual. The operations and maintenance manual communicates to facility managers the limitations that are placed upon the design. These limitations stem from decisions made during the design process. For example, heat release rates used as input data place a limitation on the use of a space. Any furnishings placed within a space that could have higher heat release rates than the heat release rates used during fire modeling could result in greater consequences than the model predicted.

The operations and maintenance manual should be written in a format that can easily be understood by people who are not fire safety professionals, since most building owners and facility managers will not have this type of background.

APPLICATION OF PERFORMANCE-BASED DESIGN

Performance-based design can be applied in one of three situations: with prescriptive regulations, with performance-based regulations, and as a stand-alone design methodology (SFPE, 2007).

Use with Prescriptive-Based Regulations

Prescriptive-based regulations provide requirements for broad classifications of buildings. These requirements are generally stated in terms of fixed values, such as maximum travel distances, minimum ratings of boundaries, and minimum features of required systems (e.g., detection, alarm, suppression, and ventilation).

Most prescriptive-based regulations contain a clause that permits the use of alternative means to meet the intent of the prescribed provisions. This provides an opportunity for a performance-based design approach. Through performance-based design, it can be demonstrated whether or not a design is satisfactory and complies with the implicit or explicit intent of the applicable regulation.

When applying performance-based design in this manner, the scope of the design is equivalency with the prescriptive provision(s) for which equivalency is sought. The intent, or performance intended by the prescriptive code provision(s), is identified to provide the goals and objectives for the design.

Use with Performance-Based Regulations

Performance-based codes and standards provide goals, objectives, and performance criteria for buildings or other structures that fall within the scope of the code or standard. Performance-based codes generally provide either specific fire scenarios that need to be addressed (as is the case in the performance-based codes developed by NFPA) or information that is intended to identify the types of fire scenarios that must be addressed (in the case of the *ICC Performance Code for Buildings and Facilities*). Performance codes may also provide additional administrative provisions, such as review or documentation requirements.

Use as a Stand-Alone Methodology

In some cases, a building owner or insurer may have additional fire safety goals beyond the minimum requirements of applicable prescriptive codes and standards. In these cases, additional or complementary fire safety goals and objectives might be identified, thus requiring additional fire protection engineering analysis and design.

For example, property protection and continuity of operations might be goals of a building owner or insurer, which might not be fully addressed in applicable regulations. The performance-based design process can be used to identify and address these additional goals.

REFERENCES

ASCE, *Standard Calculation Methods for Structural Fire Protection*, American Society of Civil Engineers, ASCE/SFPE 29-05, Reston, VA, 2005.

Custer, R., and Meacham, B., *Introduction to Performance-Based Fire Safety*, National Fire Protection Association, Quincy, MA, 1997.

ICC, *ICC Performance Code® for Buildings and Facilities*, International Code Council, Falls Church, VA, 2012.

Nelson, H., Performance-Based Fire Safety, *Proceedings of the 1996 International Conference on Performance-Based Codes and Fire Safety Design Methods*, Society of Fire Protection Engineers, Bethesda, MD, 1996.

NFPA, *Life Safety Code*, NFPA 101, National Fire Protection Association, Quincy, MA, 2012a.

NFPA, *Guide to the Fire Safety Concepts Tree*, NFPA 550, National Fire Protection Association, Quincy, MA, 2012b.

NFPA, *Building Construction and Safety Code*, NFPA 5000, National Fire Protection Association, Quincy, MA, 2012c.

SFPE, *SFPE Engineering Guide to Human Behavior in Fire*, Society of Fire Protection Engineers, Bethesda, MD, 2003.

SFPE, *SFPE Engineering Guide to Performance-Based Fire Protection*, National Fire Protection Association, Quincy, MA, 2007.

Chapter 2

Hazard and Risk

INTRODUCTION

All fires start with an event—such as an electrical failure, equipment failure, or carelessly discarded smoking materials. Once the fire starts, it can manifest itself in a number of ways: grow in size, spread to other items, activate fire detection or suppression systems, or self-extinguish. Each combination of possible sequences of actions is a fire scenario. Event trees can be used to graphically illustrate all of the possible sequences of actions following a fire event.

Figure 2.1 shows an example of an event tree for the possible course of action for a fire that starts in a room (Hui, 2006).

Following the ignition event in Figure 2.1, the smoke detector that protects the room could either activate or not activate. Similarly, the room occupants could either successfully extinguish the fire or not. If the occupants do not extinguish the fire, the sprinkler that is installed in the room could either control the fire or not. Finally, if the sprinkler is not successful, the room compartmentation could either contain the fire or it could spread beyond the room of origin. Detector operation, occupant extinguishment, sprinkler activation, and barrier containment are all subsequent events that could occur (or not occur) following the initial fire initiation event.

From the single ignition event in the room, there are eight possible scenarios that could occur. Each scenario occurs with a different probability that can be determined from the probabilities of the separate mitigation strategies occurring and Boolean algebra. If the probabilities of the mitigation strategies being successful are high, then the overall probability that the fire will not be controlled or contained within the room of origin is low.

There would likely be a number of event trees that could be prepared for the room that is illustrated in Figure 2.1. Figure 2.1 illustrates the possible scenarios that could occur following a specific ignition event, for example, a carelessly discarded cigarette. There are likely many other ignition events that could occur in this room, and an event tree could be developed for each one. In some cases, the event tree for other ignition events might be

23

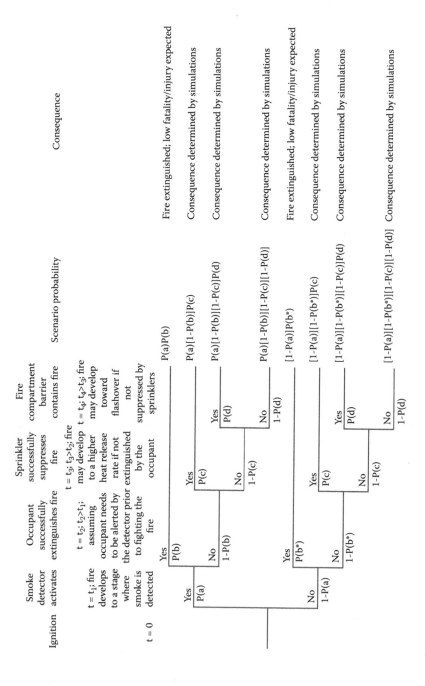

Figure 2.1 Example event tree. (Used with permission of Society of Fire Protection Engineers, copyright © 2006.)

identical or similar to Figure 2.1. Identical would mean that the same possible subsequent events could occur with the same probabilities. Similar would mean that the same subsequent events could occur with similar probabilities. However, for other ignition scenarios the event trees might be much different.

If Figure 2.1 illustrates the possible scenarios that could occur if smoking materials are carelessly discarded in a wastebasket, then other fire initiation events that start in the wastebasket might have similar or identical event trees. However, if the fire ignition event is a Christmas tree fire, then the event tree for this event would likely be much different—the occupant might not be capable of extinguishing the fire, and the probability of success of the other mitigation methods would likely be different.

Additionally, the room in Figure 2.1 would likely be contained within a larger building, meaning that the event tree could be expanded to include other events that could occur if the fire is not contained within the room of origin.

Each scenario that is a terminus on the right-hand side of an event tree represents a series of events that could occur. Each series of events occurs with a different probability, and the probabilities of some scenarios occurring are higher than others.

In performance-based design, all scenarios must be considered in some manner. There are two ways that can be used to consider the universe of possible scenarios: risk based and deterministic.

Risk-based analysis looks at the big picture of all of the possible scenarios—the consequences of each scenario are analyzed, but these consequences are weighted by the probability of the event occurring. If the total risk (which is a function of probabilities and consequences of the scenarios) is below some threshold value, then the design is considered acceptable.

In deterministic analysis, scenarios that are expected to occur with a probability above some threshold value are analyzed to determine their consequences. If the consequences of each scenario are below some threshold value, then the design is considered to be acceptable.

Overview of Hazard and Risk

The *SFPE Engineering Guide to Fire Risk Assessment* (SFPE, 2006) defines *hazard* and *risk* as follows:

> Hazard: A condition or physical situation with a potential for harm.
> Risk: The potential for realization of unwanted adverse consequences, considering scenarios and their associated frequencies or probabilities and associated consequences.

A 5-gallon container of a flammable liquid is a hazard. The flammable liquid could be ignited if the use or storage conditions of the liquid allow flammable vapors to come into contact with an ignition source. The risk posed by the 5-gallon container of a flammable liquid is a measure of how likely it is that the flammable liquid will be ignited coupled with a measure of the consequences resulting from ignition of the flammable liquid.

In the case of the 5-gallon container of a flammable liquid, there are many scenarios under which the flammable liquid could be ignited. These scenarios are dependent on the use or storage of the flammable liquid. A hazards analysis would investigate the consequences of possible scenarios. A risk analysis would analyze the probability of the scenarios occurring and their consequences. In both cases, consequences could be a measure of threat to life, property loss, business continuity, or damage to the environment.

Because conducting a risk analysis includes conducting a hazards analysis, hazards analysis could be viewed as a subset of risk analysis. However, hazards analyses are frequently performed as a stand-alone task.

The *SFPE Engineering Guide to Performance-Based Fire Protection* (SFPE, 2007) defines *risk* as the product of consequence and frequency, which would give risk units of loss per unit time. Risk can also be defined as the product of consequence and probability. Since probability is unitless, the resulting product has units of loss.

Risk is typically stated in terms of a period of time, such as the annual risk or the risk over the projected life of as building. In such cases, the probabilities used would be the probability of an event occurring within the time period, such as the probability of a fire occurring in a year.

Hazard and Risk in the Performance-Based Design Process

In the context of performance-based design, risk analysis or hazard analysis is a subset of the performance-based design process. A hazards analysis might be done to identify design fire scenarios. A risk analysis might be done to analyze the risk associated with an activity or the risk associated with a proposed design strategy. In either case, the types of hazards that would be considered would be a function of the scope of the project.

Performance-based analyses or designs may be conducted on either a risk or hazards basis. When the project is done on a risk basis, it is referred to as *risk based*. If the project is done on a hazards basis, it is referred to as *deterministic*. The key difference between deterministic and risk-based projects is how probabilities are considered.

Generally, probabilities are considered in both types of analyses; however, in risk-based analyses, probabilities are explicitly considered by including them in measures of risk. In hazard analyses, probabilities are considered by determining if a hazard is expected to occur frequently enough to warrant further analysis.

The acceptance criteria used in a performance-based design project differ depending upon whether a project is done on a risk basis or deterministic basis. For a deterministic analysis, acceptance (performance) criteria would be stated in terms of the maximum or minimum permissible effects, such as a maximum permissible upper layer temperature or heat flux resulting from a fire scenario. In a risk-based analysis, acceptance criteria would generally be in terms of acceptable loss per unit time. Note that in the case of a risk-based analysis, the loss is the summation over all scenarios identified, whereas in a deterministic analysis the effects are measured on a scenario-by-scenario basis.

Because of the difference in type of acceptance criteria and the types of calculations that are performed, whether a project will be conducted on a deterministic or probabilistic basis should be determined at the onset of the project.

With the exception of a few specialized areas (e.g., the nuclear industry) performance-based design is generally conducted on a hazard, or deterministic, basis.

HAZARD ASSESSMENT METHODOLOGY

A hazard assessment results in an identification of the scenarios of interest and their consequences. The following hazard assessment methodology is adapted from Hurley and Bukowski (2008):

1. **Define project scope.** The project scope is an identification of the problem to be analyzed. This is similar in nature to the definition of scope of a performance-based design project.
2. **Identify fire protection goals, objectives, and performance criteria.** Fire hazards analyses are generally conducted as part of a performance-based design process or to determine if the existing level of safety in a building or facility is adequate. In the case of a performance-based design, a design is prepared to achieve certain goals and objectives. In the case of a stand-alone hazards analysis, the purpose of the hazards analysis is to determine whether the hazards in a building or facility are within certain limits, where the limits are associated with specific goals and objectives. Once the goals and objectives have been developed, these will need to be quantified as performance criteria. The development of goals, objectives, and performance criteria is discussed in more detail in Chapter 1.
3. **Identify hazards.** Hazards identification can be conducted using the tools identified in the *SFPE Engineering Guide to Performance-Based Fire Protection* (SFPE, 2007): failure modes and effects analysis,

failure analysis, "what if" analysis, review of historical data and checklists, and review of statistical data. Professional judgment and experience can be used to augment these tools, and for some projects, professional judgment and experience will be sufficient. These tools are explained more completely in Chapters 1 and 3.

4. **Develop scenarios.** From the hazards that are identified, a list of scenarios can be developed. In some cases, more than one scenario may be developed for a given hazard depending upon the failure modes and contributing factors. When developing scenarios, the reliability of fire protection systems (sprinkler, alarm, compartmentation, structural fire resistance, etc.) should be considered. The development of scenarios is addressed further in Chapter 3. The number of scenarios that are developed can be reduced into a manageable set by grouping similar scenarios and eliminating scenarios that are determined to be too unlikely to merit further attention. When grouping like scenarios, a bounding scenario can be selected that is as severe as or more severe than the other scenarios in the group. The purpose in selecting bounding scenarios is that the person conducting the hazards analysis can be confident that the goals and objectives would be achieved in the group bounded by the scenario if goals and objectives are achieved in the bounding scenario. Although probabilistic aspects of fire (such as the likelihood of a scenario or the reliability of installed systems) are not explicitly considered in hazards analysis, probabilistic aspects could influence which scenarios are selected for further evaluation and which are determined to be too unlikely to merit further attention. When a scenario is eliminated on the basis that it is considered too unlikely, there is an *implied risk* that must be accepted by those who are affected. The implied risk is that if the scenario were to occur, unacceptable results may occur. The selection of scenarios is addressed further in Chapter 3.

6. **Quantify the effects of scenarios.** The effects of the scenarios that have been selected are then quantified. The quantification of scenario effects would generally be accomplished by use of fire models. The quantification method should provide output that is comparable with the performance criteria for the hazards analysis. Quantification of scenarios is discussed further in Chapter 4.

7. **Determine if goals and objectives for the hazards analysis have been achieved.** The output from the quantification of the scenario is compared to the performance criteria to determine whether the goals and objectives for the hazards analysis have been achieved.

8. **Account for uncertainty.** Uncertainty is present on most fire protection analyses, and this uncertainty must be addressed. Methods of considering uncertainty are addressed in more detail in Chapter 12. Uncertainty can be present in any of the elements described above.

Example Hazard Analysis

A storage building is located 30 m from a JP-5 storage tank. The tank is a cylinder that measures 10 m high by 15 m in diameter. The building is constructed of wood. Could the building be ignited by a fire in the storage tank?

SOLUTION

1. The project scope is to determine whether a fire in the storage tank could ignite the wood storage building.
2. The goal is property protection. The objective is to prevent ignition of the wood building from radiant heat transfer from a fire in the storage tank. From Tewarson (2008) the critical heat flux for ignition of wood is 10 kW/m². The performance criterion is that if the heat flux exceeds this value, the objective is not met.
3. The hazard is a fire in the storage tank. This will be selected as the design fire scenario.
4. From Beyler (2008) the heat flux from a pool fire is

$$\dot{q}'' = 15.4\left(\frac{L}{D}\right)^{-1.59} = 15.4\left(\frac{37.5\text{m}}{15\text{m}}\right)^{-1.59} = 3.59\text{kW/m}^2$$

   With Beyler's recommended safety factor of 2, the estimated heat flux is 7.18 kW/m².
5. The calculated heat flux is below the critical radiant flux for ignition of the wood; hence, the objective is achieved.
6. Uncertainty was considered through use of the safety factor recommended by Beyler.

RISK ASSESSMENT METHODS

As stated by the definition in the previous section, risk, in the classical sense, is the product of the potential consequences and the expected probability of occurrence of a series of scenarios. This can be stated numerically as (SFPE, 2007)

$$Risk = \sum Risk_i = \sum \left(Loss_i \bullet P_i\right)$$

where $Risk_i$ is risk associated with scenario i, $Loss_i$ is loss associated with scenario i, and P_i is probability of scenario i occurring.

Note that in deterministic analysis, scenarios might be eliminated from consideration because they are determined to be too unlikely to merit further consideration. For example, if a high-rise building is being analyzed, a scenario could be identified where a gasoline delivery truck collides with the building. In a deterministic analysis, this scenario might be considered to be too unlikely to be considered further. However, in a risk-based analysis, this scenario would need to be evaluated, although the resulting risk would be discounted by the fact that the probability of the scenario is very low.

In a risk-based analysis, a scenario could be eliminated from further consideration if no design could provide protection if the scenario were to occur. Consider, for example, a gasoline tank truck colliding with a single-family residence. It is very unlikely that any level of protection could reasonably be expected to be successful if this event were to occur.

The *SFPE Handbook of Fire Protection Engineering* classifies methods for assessing fire risk into four areas (listed in increasing level of detail) (Watts and Hall, 2008):

- Checklists
- Narratives
- Indexing
- Probabilistic methods

A checklist is a simple listing of hazards that is intended to be specific for a type or class of building. It is a risk-based method in that the types of items that are contained on the checklist are generally oriented toward conditions that contribute to fire risk. For example, flammable liquids stored in appropriate containers would give an indication of the likelihood of flammable liquids being involved in fire development. If the flammable liquids were stored in appropriate containers, the flammable liquids would likely contribute less to the overall fire risk than if they were not stored in the appropriate containers. A checklist gives a qualitative evaluation of fire risk.

A narrative is a simple listing of recommendations of things to do and things not to do. The statements on the narrative are influenced by the items that would contribute toward fire risk. As with checklists, narratives are a qualitative fire risk methodology.

Indexing involves the assignment of values (positive and negative) to fire safety features. Probabilities and consequences are considered through the values that are assigned to different features. An index method is a semi-quantitative measure of fire risk—while a numerical estimate is provided, which can be used to compare different risks, the units of the numerical estimate are not those of fire risk (such as dollars lost per year.)

Probabilistic methods explicitly consider the probability and consequence of events. Probabilistic methods include event trees, decision trees, fault trees, and influence diagrams. A less rigorous, quasi-quantitative probabilistic method is the use of fire risk matrices.

The *Guide for the Evaluation of Fire Risk Assessments* (NFPA, 2013) provides a similar classification of fire risk assessment methods. NFPA 551 categorizes risk assessment methods as qualitative, semiquantitative likelihood, semiquantitative consequence, quantitative, or cost-benefit risk.

A qualitative method treats both consequence and likelihood qualitatively. Qualitative methods would include checklists, narratives, index methods, and risk matrices. Semiqualitative likelihood methods treat likelihood quantitatively and consequences qualitatively. Examples of semiqualitative likelihood methods include actuarial/loss statistical analyses and stand-alone event tree analyses. Semiquantitative consequence methods treat consequences quantitatively and likelihood qualitatively. Examples of semiquantitative consequence methods include a fire model for which a worst credible fire scenario is selected. A quantitative method treats both probability and consequences quantitatively. A cost-benefit risk method is a probabilistic method that also analyzes the cost-benefit of different fire protection strategies.

QUANTITATIVE FIRE RISK ASSESSMENT METHODOLOGY

The *SFPE Engineering Guide to Fire Risk Assessment* (SFPE, 2006) has developed the following procedure for conducting a quantitative fire risk assessment:

1. **Define project scope.** The project scope is an identification of the problem to be analyzed. This is similar in nature to the definition of scope of a performance-based design project.
2. **Set a risk acceptability threshold.** The risk acceptability threshold is a measure of the maximum level of risk that is considered tolerable. The risk acceptability threshold is a quantification of the goals and objectives for a project in probabilistic terms. The risk acceptability threshold is typically stated as time-based rate or probability-selected consequence measures (e.g., maximum tolerable number of deaths per year or maximum tolerable probability of death in 5 years).
3. **Identify hazards.** Hazard identification is the first step in scenario development. Hazard identification involves selecting the conditions or situations with potential for undesirable consequences. Hazard analysis is explained in more detail in Chapters 1 and 3.
4. **Develop scenarios.** A fire scenario is a fire incident characterized as a sequence of events. Scenarios are developed based on the project scope and the hazards. The development of scenarios can result in a large number of options, which are refined in the next step. The development of scenarios is addressed further in Chapter 3.
5. **Select scenarios.** The range of scenarios developed must be reduced into a manageable number of scenarios that collectively represent all scenarios for further study. Scenarios can be grouped into clusters of like scenarios according to common defining characteristics. When

scenarios are clustered, a scenario that is representative of the scenario cluster can be selected for analysis.

6. **Data.** Data must be gathered as necessary to conduct probability and consequence analysis.

7. **Probability analysis.** The probability analysis determines how often the selected scenarios may be expected to occur per unit time. Where scenarios have been clustered in step 5, the probability for the scenario cluster is the sum of the probabilities of all scenarios in the cluster.

8. **Consequence analysis.** Consequence analysis is the process of determining the potential impacts of a hazard event without consideration of probability. If a scenario is selected to be representative of a cluster, its estimated consequences should be at least equal to the average of the consequences of all scenarios in the cluster.

9. **Calculate risk.** The final step in fire risk estimation is to combine the calculated probabilities and consequences into summary measures for comparison with the acceptability thresholds.

10. **Uncertainty analysis.** Uncertainty analysis involves estimating and accounting for uncertainty in consequence and probability predictions. Methods of considering uncertainty will be addressed in more detail in Chapter 12.

11. **Risk evaluation.** The calculated risk is evaluated to determine whether it is within the acceptability threshold determined in step 2.

12. **Documentation of assessment.** Following completion of the fire risk assessment, the process should be documented. The documentation is similar in nature to the documentation of a performance-based design as identified in Chapter 1.

This process is illustrated graphically in Figure 2.2.

Risk Assessment Example

A building is a fully automated warehouse that stores components used in the manufacturing of electronic equipment. A building owner's maximum fire risk exposure is determined to be $100,000 per year. Since the building is fully automated, property protection is the only goal for the project. The following scenarios are identified:

Scenario	Probability (per year)	Consequence ($)
Catastrophic failure of equipment	10^{-4}	2,000,000,000
Automation failure leading to fire	5×10^{-2}	700,000
Electrical distribution equipment failure leading to fire	5×10^{-2}	1,000,000
Rubbish fire	5×10^{-1}	500
Maintenance-initiated fires	10^{-1}	10,000

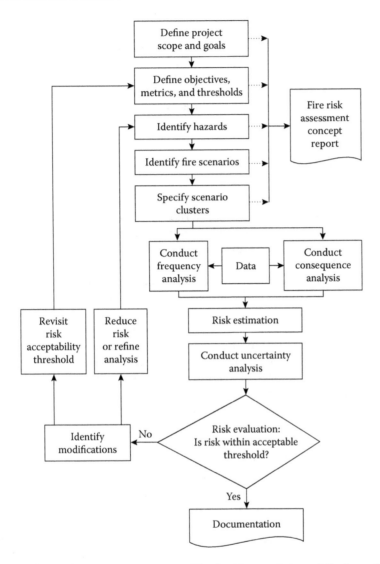

Figure 2.2 Fire risk assessment process. (Used with permission of Society of Fire Protection Engineers, copyright © 2006.)

Is the risk within tolerable limits?

In this simple example, the total risk (in 1 year) is $286,250, which is greater than the maximum threshold and is hence unacceptable. This was determined by summing the products of the probabilities and consequences for each of the scenarios. It is noteworthy that the catastrophic failure scenario accounts for the majority of the risk, even though it would occur with very low probability. In a hazard-based evaluation, this scenario might have been eliminated from further consideration because it was deemed sufficiently unlikely to merit attention.

In this example, risk reduction will be necessary to bring the risk within tolerable levels. When considering risk reduction strategies, it can be beneficial to identify the scenarios that dominate the estimated risk. Strategies that reduce the risk associated with the scenarios that dominate the risk would have the greatest impact upon reducing the overall risk.

RISK MATRICES

Another method that can be used to evaluate fire risk is a risk matrix. Because the data needed to perform a risk analysis are not always available, risk matrices are frequently used in performance-based design when it is desired to use a risk-based approach. An example risk matrix is provided in Figure 2.3.

There is a strong need for data in a fire risk assessment. Data are needed to perform probability and consequence evaluations. However, in some cases, the necessary data may be scarce, which can make it difficult to accurately estimate probabilities and consequences.

When using a risk matrix, it is not necessary to precisely estimate probabilities and consequences. Probabilities and consequences can be divided into an arbitrary number of groups, where the grouping is based on the magnitude of probability or consequence.

A matrix is constructed of the probability and consequence groupings, as shown in Figure 2.3. Each cell in the matrix is then assigned a risk categorization. Project stakeholders might then determine that certain levels of risk are acceptable and other risks are unacceptable.

When constructing risk matrices, it is necessary to ensure that the probabilities and consequences used in the groupings bound the probabilities and consequences of all scenarios. This means that the maximum consequence band must include the maximum foreseeable loss.

A fire risk matrix has the advantage that it can be applied easily or when data are not available. However, since a risk matrix does not look at the total risk associated with all scenarios, it does not give as complete a picture as a fully quantitative fire risk assessment.

Example Application of a Risk Matrix

A building is a fully automated warehouse that stores components used in the manufacturing of electronic equipment. Since the building is fully automated, property protection is the only goal for the project. Using a risk matrix, determine whether the risks are acceptable. The following scenarios are identified:

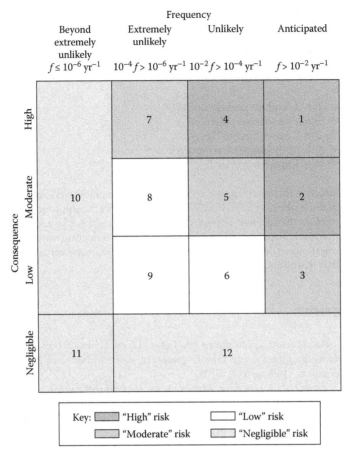

Figure 2.3 Example risk index. (Used with permission of Society of Fire Protection Engineers, copyright © 2007.)

Scenario	Probability (per year)	Consequence ($)
1. Catastrophic failure of equipment	10^{-4}	2,000,000,000
2. Automation failure leading to fire	5×10^{-2}	700,000
3. Electrical distribution equipment failure leading to fire	5×10^{-2}	1,000,000
4. Rubbish fire	5×10^{-1}	500
5. Maintenance-initiated fires	10^{-1}	10,000

Is the risk within tolerable limits?

Assume that the stakeholders are willing to accept any negligible, low, or moderate risks. Also assume that conversations with the stakeholders have developed the following consequence rankings:

High: Damage > $10 million
Moderate: $100,000 < Damage ≤ $10 million
Low: $1,000 < Damage ≤ $100,000
Negligible: Damage ≤ $1,000

The rankings for the scenarios would be as follows:

1. Bin 7 (high/extremely unlikely)
2. Bin 5 (moderate/unlikely)
3. Bin 5 (moderate/unlikely)
4. Bin 12 (negligible/anticipated)
5. Bin 3 (low/anticipated)

Based on this analysis, the risks are acceptable. However, scenario 1 is on the border between bins 7 and 4. A conservative approach would be to place the risk into bin 4, in which case it would not be acceptable. In that case, the most effective method of risk reduction would be to use methods of decreasing the probability or consequence of scenario 1 so that it would be acceptable.

REFERENCES

Beyler, C., Fire Hazard Calculations for Large, Open Hydrocarbon Fires, in *SFPE Handbook of Fire Protection Engineering*, 4th ed., National Fire Protection Association, Quincy, MA, 2008.

Hui, M.C., How Can a Fire Risk Approach Be Applied to Develop a Balanced Fire Protection Strategy, *Fire Protection Engineering*, 30, Spring 2006.

Hurley, M., and Bukowski, R., Fire Hazard Analysis Techniques, in *Fire Protection Handbook*, National Fire Protection Association, Quincy, MA, 2008.

NFPA, *Guide for the Evaluation of Fire Risk Assessments*, NFPA 551, National Fire Protection Association, Quincy, MA, 2013.

SFPE, *SFPE Engineering Guide to Fire Risk Assessment*, Society of Fire Protection Engineers, Bethesda, MD, 2006.

SFPE, *SFPE Engineering Guide to Performance-Based Fire Protection*, National Fire Protection Association, Quincy, MA, 2007.

Tewarson, A., Generation of Heat and Gaseous, Liquid and Solid Products in Fires, in *SFPE Handbook of Fire Protection Engineering*, 4th ed., National Fire Protection Association, Quincy, MA, 2008.

Watts, J., and Hall, J., Introduction to Fire Risk Analysis, in *SFPE Handbook of Fire Protection Engineering*, 4th ed., National Fire Protection Association, Quincy, MA, 2008.

Chapter 3

Design Fire Scenarios

INTRODUCTION

A performance-based design of a structure requires the evaluation of fire safety based on various possible fire scenarios. This chapter will address identifying fire scenarios. In Chapter 4, the quantification of fire scenarios as design fire curves will be covered.

The *SFPE Engineering Guide to Performance-Based Fire Protection* (SFPE, 2007) provides a two-step process for identifying design fire scenarios. As depicted in Figure 3.1, the first step is considering all possible fire scenarios that could occur in the building or portion of the building that is within the scope of the design. The second step is to reduce the population of possible fire scenarios into a manageable set of design fire scenarios. The design fire scenarios will be used to evaluate trial designs.

Determining design fire scenarios affects all facets of the performance-based analysis process. That is, occupant characteristics will affect performance criteria, and building characteristics may affect the evaluation method chosen. For example, a healthcare or other facility where occupants have health concerns will likely have lower thresholds for hazard/performance criteria. In areas where mechanical ventilation is a major issue, fire models that incorporate this feature are necessary.

In addition, the method chosen to determine fire scenarios will affect the results. It may be easier to determine fire scenarios from a survey of an existing building than from building design plans.

ELEMENTS OF FIRE SCENARIOS

Both possible fire scenarios and design fire scenarios are comprised of three sets of information: fire characteristics, building characteristics, and occupant characteristics. Fire characteristics describe the history of a fire scenario, including the first item ignited, fire growth, flashover, full development, decay, and extinction. Building characteristics describe the

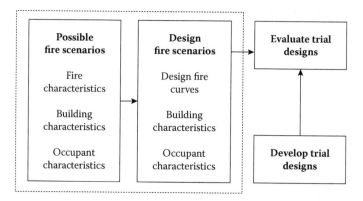

Figure 3.1 Process for identifying design fire scenarios. (From SFPE, *SFPE Engineering Guide to Performance-Based Fire Protection*, National Fire Protection Association, Quincy, MA, 2007. Used with permission of Society of Fire Protection Engineers, Copyright © 2007.)

physical features, contents, and ambient environment within the building. They can affect the evacuation of occupants, growth and spread of fire, and the movement of combustion products. Occupant characteristics determine the ability of building occupants to respond and evacuate during a fire emergency.

Fire scenarios can be developed based on the fuel loading and building conditions observed during facility surveys, based on past experience with similar buildings or by utilizing one of the methodologies discussed later in this chapter. Fire scenarios are typically postulated to represent anticipated or plausible conditions that might arise, such as temporary storage conditions. Analyses may or may not address fire scenarios involving terrorist acts, e.g., explosions, catastrophic arson fires, or other rare, but potentially severe exposures, e.g., plane crashes, depending on the intent of the analysis.

For example, if a nuclear processing facility is located near an airport, a plane crash event may be appropriate. If a facility is a known potential terrorist target, e.g., the Freedom Tower, terrorist acts may be appropriate.

The potentially applicable codes may also specify design fire scenarios. For example, the *Life Safety Code* (2012a) and NFPA 5000 (2012b) specify fire scenarios that must be addressed for performance-based designs. These fire scenarios include elements of fire characteristics, building characteristics, and occupant characteristics. However, these elements are not defined explicitly for all of the scenarios. The *ICC Performance Code for Buildings and Facilities* provides guidance on fire events in Section 1701 (ICC, 2011).

Fire Characteristics

Fire characteristics define the history of the fire itself. Fire characteristics include ignition, growth, flashover, full development, and decay. NFPA 101 and 5000 specify fire growth rates for some scenarios.

Design fire scenarios and design fire curves are discussed in Chapter 4. Building and occupant characteristics may affect the curves chosen. For example, automatic sprinklers and ventilation affect the fire curves. Availability of a fire brigade or occupants who are trained to address fire hazards may also impact fire growth.

Building Characteristics

Building characteristics describe the physical features, contents, and ambient environment of the building. Building characteristics can be grouped into the following categories:

1. Architectural features
2. Fire load
3. Structural components
4. Fire protection systems
5. Building services and processes
6. Operational characteristics
7. Fire department response characteristics
8. Environmental factors

These factors can affect the evacuation of occupants, growth and spread of fire, and movement of effluent gases.

For many projects, it is difficult to control every item on the list. It may not be possible to stipulate every conceivable detail about building operations and architectural features because other professionals, like architects or engineers of other disciplines, will be involved with designing some of these as well. Furthermore, some factors are out of the designer's control, such as the weather outside of the building. In these cases, it will suffice to identify reasonable bounding values for the features and utilize them in developing design fire scenarios. Other members of the design team, such as the architect, mechanical and electrical engineers, and enforcement officials, can be used as resources relative to building features, operational trends, or design weather conditions.

Architectural Features

The architectural features represent the composition of a building. Architectural features affect egress routes, smoke transport, and fuel loading. The architectural features include:

1. The size and location of exits. A factor to most analyses is how long it takes the occupants to get out of the structure. The size, location, and travel distance to reach the exits all affect the evacuation time.

2. Room area, geometry, ceiling height, and ceiling configuration (e.g., sloped or use of exposed beams). Examples include large open spaces that allow for smoke to gather or may result in smoke movement to other areas of the building. Sloped or obstructed ceilings affect predictions of hazards, activation of detection and suppression systems, and smoke layer depth. Each of these issues potentially requires a more complex analysis technique.

3. Interior finish flammability and thermal properties. Combustible interior finishes (e.g., wood) increase the potential for fire spread and flashover as well as fire severity. If flashover occurs, combustible interior finishes contribute to the fire load. For example, a wood interior finish may be used around an entire theater. If seating is located adjacent to the wood, the fire may spread and involve the wall, changing the predicted scenario.

4. Construction materials, such as walls, partitions, floors, and ceilings. These affect the ability of the building components to act as a barrier to flame spread and heat conduction. Fire-rated partitions provide resistance to loss of integrity and temperature transmission. Smoke partitions limit smoke movement.

5. Position, size, and quantity of openings or areas of low fire resistance (e.g., windows and doors). These can affect the size of fires by providing ventilation. The size of potential natural ventilation paths can be determined from estimates of opening sizes, assuming that all of the openings are open. Window breakage can be also estimated utilizing computer programs.

6. Configuration and location of voids. Void spaces can allow fire and smoke to spread throughout a building. The presence of combustible linings in void spaces, e.g., wood or some expanded foam insulations, could contribute to fire spread.

7. Number of stories above and below grade. The building height affects stack effect, the number of people within the building, and their egress times, as well as the ability for mobility-impaired persons to evacuate alone. In a code compliance approach, the number of levels does not directly affect the stair width, whereas in a performance-based design, the number of levels directly affects the egress time.

8. Location of the building on the site in relation to property lines and other buildings or fire hazards. The building location affects the ability of fire to spread to or from adjacent properties and the likelihood of igniting material outside of the building.

9. Interconnections between compartments. Connections between compartments affect the spread of smoke and flame throughout the

building. Potential interconnections include windows, doors, pass-through slots, atriums, ducts, and air transfer grills.

10. Proximity of hazards to vulnerable points. The proximity of hazards to items that are critical can determine how severely the mission of a building is impacted by a fire event.

Fire Load

Fire load (also sometimes called fuel load) consists of different terms used to characterize the combustible items that could burn. The fire load of a space may be expressed differently depending on the need. It may be expressed as specific fuel packages or as the mass of combustible objects per unit area of floor space. The designer should set a definition and use it consistently throughout the project.

Fire load usually includes furnishings, office supplies, flammable liquids, and other items that could be encountered within the space currently, occasionally, or at some future time. Although the fire load is typically tailored to the occupancy envisioned by the designer or owner, unless the occupancy is tightly controlled, fire loads should be expected to fluctuate throughout the lifetime of a building.

Despite the foreknowledge that loads will change, it is difficult to exactly predict how much. Therefore, some allowance or safety factor should be applied to the design. Conservative estimates (i.e., typically tending toward higher loads) should be used in the design scenario as well to offer an extra margin of safety.

The fire load that is selected will govern the use of the structure, since exceeding the design fire load could result in a more severe fire. Therefore, reasonable limitations should be established that are acceptable to the stakeholders, and reasonable means should be provided to safeguard against exceeding the limits.

Structural Components

Structural components are the elements of a building that support the weight of the contents and the building itself. Examples include steel beams, concrete columns, and precast floor slabs. Factors relevant to design fire scenarios include (SFPE, 2007):

1. Location and size of load-bearing elements. The location and size can create vulnerabilities for collapse or guide fuel package positioning. Beams are potentially exposed to fire plumes or a hot upper layer.

Columns are potentially exposed to local fires. Both beams and columns would be exposed in a flashover scenario.

2. Construction materials of structural elements. The material properties of structural elements directly affect the fire resistance and endurance of structural members.

3. Protection materials. The characteristics of protection materials (e.g., insulation) can lengthen the expected life span of components in fire.

4. Design structural loads. The design structural loads are important because heavily loaded members could fail earlier than lightly loaded members.

Fire insult to structural members is a key consideration in building design and fire protection engineering. Depending on the goals, protecting the occupants of a building could involve designing so that the building remains standing indefinitely, or at least long enough to evacuate or defend in place. Structural components and fire resistance are discussed in Chapter 9.

Fire Protection Systems

Fire protection systems encompass the active and passive components used to protect against fire. These include:

1. Detection systems, such as smoke, flame, and heat detection
2. Notification systems, including strobes, horns, and voice systems
3. Suppression systems, including various media (e.g., water, carbon dioxide, foams, etc.)
4. Smoke control to limit fire product spread or exhaust fire products
5. Fire-resistant compartmentation
6. Smoke barriers

The fire protection systems may already be assumed in place, such as in an existing building or if all parties agree a priori that a sprinkler system will be provided. Alternatively, the design may be focused on creating these systems. In these latter cases, preliminary system designs should be used by the analysis, and each iteration of the process should refine the design of the fire protection systems. After a number of trials, a system should exist on paper that satisfies the performance criteria and which can be included in the final design. This is discussed in more detail in later chapters.

Building Services and Processes

Building services and processes refers to the ventilation, electrical, and plumbing equipment present within a building. The ventilation equipment and natural ventilation present have an effect on the ambient environment and smoke transport throughout a space, as well as potentially limiting the oxygen available to the fire and the resulting size of the fire. Mechanical ventilation can be determined from site surveys, building mechanical plans, or testing and balancing reports.

The electrical system and distribution equipment can create vulnerabilities when central panels are destroyed by fire. Building services also represent potential ignition sources such as furnaces or incinerators.

A fire protection engineer may have limited influence on the design of building services and processes in some projects. However, input is important to influence fire safety aspects of critical services. That is, automatic control of mechanical ventilation systems can influence stack effect-induced flows. Furthermore, even though electrical wires will be run because electricity is required by virtually all buildings, the engineer can request alternate routing paths, particular cable types, and fire-stop sealant materials for through penetrations.

Operational Characteristics

The operational characteristics refer to the level of activity in the building as a function of time and day of the week. The designer should determine the activities that occur and the distribution of people performing them. Knowledge about the particular occupant load present in a space helps to refine the egress analysis and detail the hazards associated with these times. For example, in a fully occupied office building, the hazards may be reduced since a fire may be identified earlier by occupants. However, the amount of time to evacuate may be increased.

Fire Department Response Characteristics

Fire department response may or may not be incorporated in a performance-based analysis. Typically, the uncertainty associated with fire department response limits their impact in all but situations where a trained brigade is on site. Response by the fire department relates primarily to the capability of the fire department to arrive with adequate personnel in adequate time to suppress or control a fire. The elements of fire department response include:

1. Response time, or the time between notification of the brigade and its arrival. (A second factor here is the delay between ignition and local detection.)
2. Availability of the fire department to respond. This is influenced by whether the fire department is in the station or on another call.
3. Accessibility of fire appliances to the building site. It will be necessary for the fire department to maneuver equipment to the building location.

4. Access within the building. Once the fire department is on site, they will need access to the fire location. This is influenced by stairways and corridors and impediments such as fences and gates. Accessibility to standpipes can also be a factor when fighting fires within structures.

A structure or industrial machinery may be salvaged if the fire brigade arrives quickly and acts appropriately. In addition, fire spread to other structures or wild land exposures may be addressed by fire department response.

Fire department operations may also affect the goals of the analysis. If fire department response is a goal that must be addressed, structural stability must be assured for the length of their operations.

Environmental Characteristics

Environmental factors inside and outside the room of origin and building should be considered. For example, the interior air temperature gradient may prevent smoke from rising to the ceiling in tall spaces such as an atrium, rendering smoke detectors there unable to respond to a developing fire. Exterior wind currents may induce building flow path effects. Also important are ambient sound levels, which can affect alarm audibility.

Environmental conditions may affect designs differently depending on the project's geographical location. Stack effect will be different for a building that is in Alaska than it is for a building in Dubai. Because of wind effects, a smoke control design in San Francisco will be different if the building has operable windows.

Occupant Characteristics

Occupant characteristics are discussed in detail in Chapter 6. Several elements are critical to fire scenario development. These issues include:

1. Number and distribution of occupants
2. Occupant health characteristics
3. Occupant response characteristics

The number and distribution of occupants directly affects the ability and time to evacuate. Five thousand people located in a single-level convention center with code-compliant exits will evacuate differently than 5,000 in a 20-story office building with code-compliant exits.

Occupant health characteristics affect the movement speed and performance criteria used to address hazards to the occupants. Disabled occupants will travel at slower speeds and need to rest more often than able-bodied occupants. In addition, they may be more susceptible to fire product exposures such as carbon monoxide due to breathing concerns. Chapter 6 provides estimates of egress speed and the percentage of disabled occupants in the population.

Some facilities are typically only occupied by able-bodied personnel. For example, an airport traffic control tower requires occupants to be capable of movement. However, the occupants may become temporarily mobility impaired due to injury or illness.

Human reaction requires analysis and incorporation into performance-based designs. Decision making is included in some evacuation models. People require time to decide to evacuate in response to alarms and also perform activities before they begin to move. These activities include saving and collecting property, searching for others, or planning escape routes.

METHODS TO DETERMINE POSSIBLE FIRE SCENARIOS

For many situations, the possible fire scenarios are obvious. For example, storage tank farm scenarios may include the ignition of one of the tanks or a boiling liquid expanding vapor explosion (BLEVE) in case pressure regulators fail. For other situations, the fire scenarios are not as obvious.

The *SFPE Engineering Guide to Performance-Based Fire Protection* (SFPE, 2007) identifies a number of analysis techniques that can be used to identify possible scenarios. These techniques include:

- Failure modes and effects analysis (FMEA), where the different types of failures that could occur are studied and the effects of those failures are analyzed (Stamatis, 2003). FMEA is an analytical tool used in mechanical engineering design and manufacturing management, but can be adapted for fire protection purposes as well. The purpose of FMEA is to generate a list of failure scenarios and rank them in the order in which they should be addressed.
- Failure analysis, where potential causes of failures are identified and the expected system performance is investigated (Marcel Dekker, 2004).
- "What if" analysis, where expert opinion is used to consider possible events and the consequences of those events (Center for Chemical Process Safety, 1992). The simplest type of analysis poses the question "What if...?" The engineer imagines possible fuel loading and ignition scenarios and ponders the consequences and likely growth pattern should it occur.
- Hazards and operability studies (HAZOPS), where facility operations are studied to identify concerns (Hyatt, 2003). The HAZOPS methodology is often applied to the operation and design of manufacturing

plants to determine manners in which hazards develop and identify operability problems. For the purposes of planning design fires, the goals of HAZOP studies are manipulated slightly to become more relevant. The engineer can use HAZOP studies to envision possible hazards (fuel spills, ignitions, unsafe fuel loading conditions) and the operational mistakes that can cause or exacerbate them.

- Historical data, manuals, and checklists, where past events in the building or a similar building are studied to consider whether similar events could occur in the building that is being designed or modified. Manuals and checklists can be studied to consider warnings, cautions, or operational sequences that could lead to a fire if not followed.
- Statistical data of fires across broad classifications of buildings.
- Other analysis methods, such as preliminary hazard analysis, fault tree analysis, event tree analysis, cause-consequence analysis, and reliability analysis.

The results of FMEA, "what if," HAZOPS, and other analysis techniques can produce a large number of fire scenarios. Frequently, the number fire scenarios can be reduced to a small group of bounding scenarios through the use of engineering judgment. Correlations and other back-of-the-envelope-type calculations, as well as subsequent results from other fire scenarios, can be used to augment the use of engineering judgment. There are also probabilistic approaches that can be used to reduce the number of fire scenarios. These probabilistic approaches expand on techniques used in the determination of fire scenarios. Consultation with the enforcement official can also provide insight into good choices for fire scenarios.

As part of the scenario development in existing facilities, an important tool that can be utilized is a tour of the facility. The primary purpose is to collect the information necessary to determine fire scenarios as well as relevant calculation parameters. It is useful to verify or clarify any previously provided information.

The tour provides an opportunity to notice problems or inconsistencies that would affect the analysis. The major shortcoming of a tour is that it is a snapshot in time; therefore, it is not useful for determining variations that may occur over time.

For new buildings, a tour of similar facilities is useful. For example, walking through a laboratory that performs similar research to the one being evaluated can provide insight into what may occur. In some cases, however, the possible fire scenarios must be built on past experiences, statistical data, or other materials. Examples of possible fire scenarios will range from the mundane to the outlandish. Table 3.1 lists potential fire scenarios for a building.

Table 3.1 Potential Fire Scenarios

Design Fires	Building Characteristics	Occupant Characteristics
Match	Central atrium	Unoccupied
Trash can	Combustible timber members	Occupied by an expected number of
Computer	Wet-pipe sprinkler system	healthy people
Desk	throughout	Occupied by a code-bound maximum
Arson	Mechanical ventilation	number of healthy people
Flammable	provided on each story	Occupied 50% by disabled people
liquid spill	through central supply duct	Occupied mostly by infants and some
Airplane crash	Unoccupied on nights and	adults
	weekend	Occupied by "commandable" military
	Far from fire station	personnel

An example of a single fire scenario would include all of the following:

1. Design fire. A flammable liquid fire located at the main entrance of the facility. Ignition is an arson fire. Fast growth with ignition of surrounding combustibles.
2. Building characteristics:
   a. Architectural features. Large central atrium, noncombustible or limited-combustible interior finishes.
   b. Fire load. Typical office occupancy.
   c. Structural components. Concrete beams and columns.
   d. Fire protection systems. As a planned building, no fire protection systems are yet in place.
   e. Building services and processes. HVAC forced air system. No operable windows.
   f. Operational characteristics. Building occupied 24 h per day.
   g. Fire department response characteristics. Volunteer fire department 5 km away.
   h. Environmental factors. Building in Washington, DC.
3. Occupant characteristics:
   a. Building is loaded at maximum occupancy.
   b. Occupants are awake and alert.
   c. Use of structure mandates occupants are mobile, but other capabilities (e.g., sight and hearing) mirror the general public.

DESIGN FIRE SCENARIOS

Given the large number of possible fire scenarios for any given performance-based design project, it is usually necessary to reduce the possible fire scenario population to a manageable number of design fire scenarios that will be used for evaluating trial designs. For most design projects, this can be accomplished in part by excluding scenarios that are unlikely to occur or that would result in an acceptable outcome regardless of the trial design strategy that is used. However, for a fire scenario to be excluded from further analysis because it is considered too unlikely, all stakeholders must recognize and accept that if the scenario were to occur, an unacceptable outcome may result.

The determination of design fire scenarios can be done in several ways. In most analyses, the fire protection engineer will propose to the owner and code official the appropriate fire scenarios to be evaluated. The code official will then concur or offer suggestions on what should be included. The owner may also provide input, but typically he or she leaves scenario selection to the fire protection engineer.

The fire protection engineer must ensure the owner understands the implications of the scenarios selected, especially if it puts limitations on the use or occupancy of the building. If scenario selection limits the use of the atrium or floor of a gymnasium, the owner may lose revenue, and therefore would have a significant interest.

In other cases, scenario development can be based on the following:

- All of the project stakeholders meeting and discussing what they think should be analyzed, based on past protocol in a jurisdiction.
- In some jurisdictions, performance-based design is performed more often than others, e.g., in Las Vegas or federal government complexes such as the U.S. Capitol. In these cases, the code official might have certain fire scenarios that must be incorporated into the analysis. Therefore, the stakeholders should be asked whether they want specific scenarios included.
- The applicable code may identify the design fire scenarios, such as NFPA 101 (2012a) or NFPA 5000 (2012b). This is discussed in more detail later in this chapter. Even when such a code is applicable to a design, the fire protection engineer should work with project stakeholders to determine if there are other scenarios that should be considered.

Another method of reducing the number of fire scenarios is to select bounding scenarios, wherein if the performance criteria can be achieved in these scenarios, it can be safely assumed that they would be achieved in the scenarios that are not specifically considered. In addition, fire scenarios may be based on issues of noncompliance with prescriptive requirements. For example, if there are travel distance issues in a particular room, only fire scenarios that would impact occupant egress may be selected.

LIFE SAFETY CODE **REQUIREMENTS**

For a performance-based assessment, the *Life Safety Code* requires the assessment of specific fire scenarios to determine the overall life safety capabilities of the building (NFPA, 2012a). A summary of required *Life Safety Code* and NFPA 5000 (2012b) scenarios is as follows:

1. Typical occupancy-specific design fire scenario
2. Ultra-fast developing fire in the primary means of egress
3. Fire in an unoccupied room near a high-occupancy space
4. Concealed space fire near a high-occupancy space
5. Slow developing shielded fire near a high-occupancy space
6. Most severe fire associated with the greatest fuel load
7. Outside exposure fire

In jurisdictions where NFPA 101 or 5000 is applicable, the analysis must include at least a discussion of all of these scenarios. A detailed analysis of each specific scenario may not be appropriate for each building. For example, a concealed fire may not be feasible in some noncombustible construction and then can be dismissed.

In addition, the analysis of each scenario may not be required to demonstrate that a potential problem does not exist. Design fire scenarios are developed to identify severe exposures. Disabling a fire protection feature will reduce the level of safety. A code official may accept a scenario that results in a hazardous condition if the reduced level of safety and potential hazard is quantified and appears unlikely.

The *Life Safety Code* (NFPA, 2012a) stresses the importance of choosing realistic scenarios; that is, scenarios must not be trivially easy or hopelessly severe. Rather, the proposed scenarios should be as challenging as what could occur in the building.

Life Safety Code **Design Fire Scenario 1**

Life Safety Code fire scenario 1 is an occupancy-specific fire scenario representative of a typical fire for the occupancy. For a mixed use facility, several fire scenarios are needed to represent the various occupancy uses. In fact, scenario 1 tends to include multiple subscenarios, even for single-use spaces, because many different common fires can occur. The *Life Safety Code* requires that occupant activities, number and location, room size, furnishings and contents, fuel properties and ignition sources, ventilation conditions, and the first item ignited be identified in the analysis. The purpose of scenario 1 is to ensure that the statistically most likely type of fire is considered in a design scenario (Cote and Harrington, 2012).

Life Safety Code Design Fire Scenario 2

Life Safety Code fire scenario 2 is an ultra-fast developing fire in the primary means of egress with interior doors open at the start of the fire. This design scenario addresses the concern regarding a reduction in the number of available exits and the spread of fire effects. The fundamental question being answered regarding the spread of fire effects is "What is the maximum extent of smoke that may be experienced if an egress path is blocked?" (Cote and Harrington, 2012).

Typically, the amount of combustibles located in a means of egress is minimal. Therefore, the focus is on transient or temporary fuel loadings. An arson fire, a janitorial cart full of solvent-soaked rags, or transient construction materials deliberately ignited may be reasonable for some spaces. The materials expressly block an egress path.

Life Safety Code Design Fire Scenario 3

Life Safety Code fire scenario 3 is a fire that starts in a normally unoccupied room that can potentially endanger a large number of occupants in a large room or other area. This scenario addresses the concern regarding a fire starting in a normally unoccupied room and migrating into the space that can potentially hold the greatest number of occupants in the building.

The classic example for scenario 3 is a storeroom adjacent to a dining or dancing hall. Stored linens, napkins, liquor, etc., can be ignited by an electrical source or covert smoking activities. In addition, occupants may be impaired due to drinking or distracted due to the event that is occurring.

Life Safety Code Design Fire Scenario 4

Life Safety Code fire scenario 4 is a fire that originates in a concealed wall or ceiling space adjacent to a large, occupied room. This design fire scenario addresses the concern regarding a fire originating in a concealed space that does not have either a detection system or suppression system and then spreads into the space within the building that can potentially hold the greatest number of occupants.

For a business occupancy, an electrical fire in the combustible wall of a conference room could be used as this scenario. In a museum, many walls include wood to allow for the collections to be displayed and supported by the wall. Therefore, a fire involving the exposed combustibles in the interior of the partition exposing a gallery is a potential fire scenario.

Life Safety Code Design Fire Scenario 5

Life Safety Code fire scenario 5 is a slowly developing fire, shielded from fire protection systems, in close proximity to a high-occupancy area. This design fire scenario addresses the concern regarding a relatively small ignition source causing a significant fire.

An example of this scenario is a cigarette fire in a trash can or laundry basket. The *Life Safety Code* recommends that if people are expected to be mobility impaired, the room of origin is selected to be a room full of such individuals. If mobility is expected to be normal, the trash can is placed in a place of assembly or area characteristic to the use of the property so that it is shielded from suppression systems.

Life Safety Code Design Fire Scenario 6

Life Safety Code fire scenario 6 is the most severe fire resulting from the largest possible fuel load characteristic of the normal operation of the building. This design fire scenario addresses the concern regarding a large, rapidly developing fire with occupants present.

This scenario is usually the most challenging one considered by the designer. It includes the largest fuel load possible in the normal operation of the space. This load may be a sectional sofa in a residence, bedding or medical equipment in a healthcare facility, or pool chemicals (oxidizers) in a big box store. This scenario should include exposures that are introduced by operations, including flammable liquids and gases for powering forklifts.

Life Safety Code Design Fire Scenario 7

Life Safety Code fire scenario 7 is an outside exposure fire. This scenario addresses the concern regarding a fire starting at a location remote from the area being considered and either spreading into the area, blocking escape from the area, or developing untenable conditions.

Some possible examples include fires from landscaping shrubs, encroaching forest fires, and fuel storage tank fires. Additionally, contents in the building may ignite via radiant heat flux through windows.

Life Safety Code Design Fire Scenario 8

Life Safety Code fire scenario 8 is a fire originating in ordinary combustibles in a room or area with each passive or active fire protection system independently rendered ineffective. This set of design fire scenarios addresses concerns regarding each fire protection system or fire protection feature, considered individually, being unavailable. The *Life Safety Code* (NFPA, 2012a) mentions examples of unprotected openings between floors or between fire walls or fire barrier walls, failure of fire doors to close automatically, shutoff of sprinkler system water supply, inoperative fire alarm system, inoperable smoke management system and automatic smoke dampers blocked open.

This scenario is not meant to utilize the highest imaginable fuel loading as in scenario 6. Rather, a normal amount of combustibles is anticipated.

The code provides an exemption for components with known high reliability (e.g., concrete beam encasement) or acceptable performance in the absence of the system. To obtain this exemption, the code official will typically require an analysis of system reliability and the performance without the system. However, the system does not have to meet stated performance goals and objectives.

SUMMARY

Design fire scenarios are used as input for the evaluation of the fire hazard or risk associated with a facility. Design fire scenarios are a subset of possible fire scenarios. Fire characteristics, building characteristics, and occupant characteristics are all necessary elements. Scenario development should address stakeholders' input and applicable code criteria.

REFERENCES

Cote, R. and Harrington, G.E., eds., *Life Safety Code Handbook*, National Fire Protection Association, Quincy, MA, 2012.

Center for Chemical Process Safety/American Institute of Chemical Engineers, *Guidelines for Hazard Evaluation Procedures, with Worked Examples*, Center for Chemical Process Safety/American Institute of Chemical Engineers, New York, 1992.

Hyatt, N., *Guidelines for Process Hazards Analysis (PHA, HAZOP), Hazards Identification, and Risk Analysis*, Dyadem Press, Richmond Hill, Ontario, 2003.

ICC, *International Code Council Performance Code for Buildings and Facilities*, International Code Council, Washington, DC, April, 2011.

Marcel Dekker, *Practical Engineering Failure Analysis (Mechanical Engineering)*, Marcel Dekker Press, New York, 2004.

NFPA, *Life Safety Code*, NFPA 101, National Fire Protection Association, Quincy, MA, 2012a.

NFPA, *Building Construction and Safety Code*, NFPA 5000, National Fire Protection Association, Quincy, MA, 2012b.

SFPE, *SFPE Engineering Guide to Performance-Based Fire Protection*, National Fire Protection Association, Quincy, MA, 2007.

Stamatis, D., *Failure Mode and Effect Analysis: FMEA from Theory to Execution*, ASQ Quality Press, Milwaukee, WI, 2003.

Chapter 4

Design Fires

INTRODUCTION

Fire protection engineering involves predicting the development and consequences of fire. To predict fire development, it is necessary to be conversant in the nature of flame spread and burning rates of a variety of materials and items in multiple configurations and arrangements.

For most predictions of the consequences of fire, the development of the fire itself is used as an input. The input data are usually in the form of heat release rate versus time. While there are models to calculate or simulate the development of a fire, they have limited applicability and should not be relied upon absent the fundamental knowledge and application of fire growth and burning information. This chapter will present a methodology for predicting the development of a fire for use in determining the consequences of fire. This process is known as developing a design fire or simply developing a fire curve.

Design fires need not be exact and are not precise predictions of what will happen in a fire. Design fires are meant to be a representation of anticipated fires. Limitations of current modeling technology and data make it impossible to create exact predictions of how a potential fire will burn.

A general methodology for development of design fire curves is as follows:

1. Define the fire scenario. Definition of the fire scenario is discussed at length in Chapter 3.
2. Determine the first item ignited. Typically, the first item to be ignited is chosen for each scenario. That is, there can be many first items ignited because there are multiple scenarios to consider. The first items ignited are chosen based on known or assumed fuel packages and the probability that the item would ignite.
3. Develop a design fire curve for first item ignited. Development of the design fire curve is the principal topic of this chapter.
4. Modify the design fire curve based on enclosure effects. Enclosure effects include radiative feedback, ventilation limited burning,

suppression activities, etc. Fire is a dynamic phenomenon influenced by changes in air, fuel, and heat. Modification of any of these factors can increase or decrease the size of the fire.

5. Determine if additional items are ignited.
6. Adjust design fire curve based on additional items ignited.
7. Repeat steps 4 to 6 as necessary.

DESIGN FIRE CURVE

A design fire curve is a description of the development of a fire for a particular set of conditions. Typically, this design fire curve will be described in terms of heat release rate versus time, but the design fire can also be described in other ways, such as a set of probabilities, temperatures, or smoke/toxic emissions production. The exact method used will depend on the type of analysis. For example, an office fire would be expressed as a heat release rate curve for determining sprinkler activation, a set of probabilities for a risk-based analysis, a set of temperatures for fire barrier analysis, or smoke productions rates for smoke control work. For most fires, the design fire curve is broken down into four main sections (see Figure 4.1): ignition, growth, fully developed, and decay. Each of these sections represents one particular aspect of the fire.

Ignition

Ignition is the point at which fuel, oxygen, and heat combine to begin combustion. During this stage, the heat release rate is negligible. The fire can either self-extinguish or transition to the next stage. It can also remain at

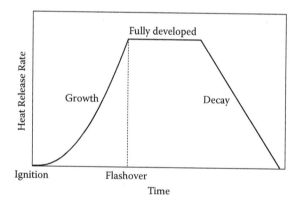

Figure 4.1 Four main sections of a fire curve. (From Hurley and Rosenbaum, 2008. Used with permission of Society of Fire Protection Engineers, copyright © 2008.)

this stage for an extended period of time. Smoldering and incipient burning would be included in this stage due to the negligible heat release rates.

For most deterministic approaches, ignition is just assumed to occur. This is due to the availability of fuel (combustibles), oxygen (air), and ignition sources (e.g., electrical, cutting/welding, arson) for a particular situation, as well as the negligible impact the ignition phase tends to have on the analysis. It should be noted that the ignition phase is not always negligible. Long, smoldering fires can cause fatalities or damage to properties due to the release of toxic gases, even if the fire self-extinguishes.

For a probabilistic approach, ignition is an important factor in the design of a fire because it helps establish the importance of a particular scenario. Having a low probability of ignition does not rule out a particular scenario. However, the consequences of the other sections of a design fire may be so large that even cases having low probability of ignition need to be considered. That is, the risk remains high and the scenario retains its importance.

Even though ignition is often assumed as the start of the design fire curve, it can play an important role in the development and growth of the fire. Ignition characteristics will typically determine the fire growth to an adjacent object. Additionally, the source of ignition can affect the fire curve. For example, an ignition source that preheats a large amount of combustibles, e.g., an arson exposure, will generate a faster fire growth than that of a cigarette.

Ignition characteristics can also be used to help identify potential fire scenarios. Objects that are readily ignitable, e.g., gasoline pools, are indicators of likely fires. Conversely, knowing that a major ignition source is present, such as welding, can be helpful in determining potential scenarios for combustibles in close proximity.

Growth Phase

The section of the fire curve following ignition is the growth phase. This phase begins when the fire exhibits a heat release rate that increases beyond a negligible level. Flame spread is the typical mechanism that governs fire growth. Flame spread is the phenomenon of flames moving across an object, igniting that which with they come into contact. The flame spread rates depend on the state of matter (solid, liquid, gas) as well as the type of fuel, configuration/orientation of the fuel, and airflow direction. Table 4.1 gives some relative flame spread rates for common situations.

Fire growth rates can be characterized if the flame spread across an item or pool surface is known. For example, the flame spread rates across liquid fuel fires are characterized by Gottuk and White (2008) and can be directly related to the increasing rate of heat release. Knowing the fuel characteristics allows for estimation of the flame spread and then a calculation of the growth rate.

Table 4.1 Relative Flame Spread Rates (Order of Magnitude Estimate)

Phenomenon	Rate (cm/s)
Smoldering	10^{-3} to 10^{-2}
Lateral or opposed flow or downward spread on thick solids	$\sim 10^{-1}$
Upward or wind-aided spread on thick solids	1 to 10^2
Horizontal spread on liquids	1 to 10^2
Forest and urban fire spread	1 to 10^2
Premixed flame speeds	
Laminar deflagration	10 to 10^2
Detonation	$\sim 3 \times 10^5$

Fire growth is also influenced by feedback of heat from the flames to the surface of the burning object itself. Objects placed in a corner or along a wall will typically grow faster than the same object located in the center of a room, away from walls. This is due to radiative feedback from the fire back to itself via reflection from adjacent walls or surfaces (Torero, 2008; Hasemi, 2008; Lattimer, 2008). Figure 4.2 shows an example of an enclosure-increasing fire growth rate due to radiative feedback. The effect of radiative feedback on the heat release rate of an object is calculated by some fire models. For most models it must be separately adjusted as part of the input.

In addition to direct flame spread, fires can also grow indirectly through the ignition of secondary fuels by conduction, convection, or radiation. As a fire grows and ignites secondary fuel packages, the growth curve will represent the sum of the contributions to the heat release rate from each item. The character, quantity, and location of secondary fuel packages must be analyzed to determine if they will actually become involved.

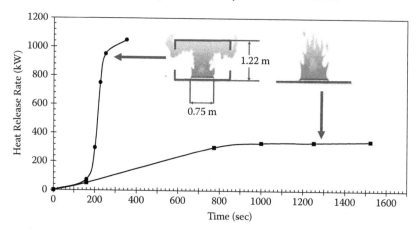

Figure 4.2 Effect of enclosure (0.76 × 0.76 m PMMA slab). (From Drysdale, D., *An Introduction to Fire Dynamics*, 3rd ed., John Wiley, New York, 2011.)

Under certain circumstances, the fire can grow rapidly from indirect ignition of secondary fuels. As a fire grows, the temperature in a compartment rises. These temperatures can rise to such a point that all of the nonburning combustibles in the room ignite almost simultaneously. This rapid ignition is known as flashover, and flashover typically occurs with upper layer temperatures between 500 and 600°C (Walton and Thomas, 2008). Flashover marks a transition point between a growing fire and a fully developed fire. Not all fires progress to flashover.

The growth period of a fire ends when it becomes either fuel or ventilation controlled. Fire growth in a fuel-controlled fire ceases when the fire has grown to its maximum size and is not able to ignite additional fuels or the fuel supply is depleted. Fire growth with a ventilation-controlled fire ends when it reaches a size governed by available oxygen supply. Rapidly growing fires may initially exceed the maximum heat release rate dictated by the ventilation conditions by consuming the oxygen initially present in the space. These fires will eventually transition back to the ventilation-controlled level when the excess oxygen is depleted.

Fully Developed

Once the fire stops growing, it transitions to the third stage of a fire curve: the fully developed region. At this point, the heat release rate plateaus and will remain at this level until the fire grows again or starts to decay. How long it remains at this level depends on the amount of fuel remaining, ignition of other fuels, ventilation conditions, and initiation of manual or automatic suppression activities.

The heat release rate of a fire at the fully developed stage can be determined based on large-scale test data, calculated based on the surface area of the burning fuel, calculated based on the amount of ventilation in a room, or estimated based on the predicted activation of suppression systems.

Large-scale tests potentially provide a direct indication of maximum heat release rates for certain scenarios. For example, Babrauskas (2008) and NFPA 72 Appendix B (2013) provide several methods to predict fully developed heat release rates for a variety of materials based on testing.

For fuel-controlled fires, the fully developed stage typically depends on the exposed surface area. For example, the peak heat release rate of a liquid pool fire is based on the area of the pool. Unfortunately, surface areas of liquid pool fires depend not only on the quantity, but also on the spill depth. This spill depth depends on confinement and flatness of the surface as well as the viscosity of the liquid. Gottuk and White (2008) provide methods to determine the surface area for liquid pool fires.

Estimating the surface area for fuel-controlled fires with combustibles that are relatively thin compared to their lengths can be more complex. These fuels may burn out in longer burning portions before the flame spreads across the entire surface. Cables are a good example of this phenomenon, where the ignition point may burn out prior to the full cable being ignited. In this case, the maximum fire size of the fully developed stage is based on the amount of material burning between the two edges of the flame.

The peak heat release rate may be governed by the amount of oxygen available in and flowing into the room. Calculation methods are available for ventilation-controlled fires (SFPE, 2004). Under steady-state conditions, the size of the fire is related to the combination of mechanical and natural ventilation present. A larger size fire may be supported early in the fire due to oxygen contained in the room.

Many methodologies also implement the concept of a limiting oxygen index or concentration. This value represents a lower combustion limit, below which combustion will not be sustained. For many combustibles, the limiting oxygen index is around 12.5 to 14.5% (Beyler, 2008b) but decreases with increasing compartment temperatures.

Suppression system activation usually ceases the growth of a fire. If suppression is inadequate, the fire will continue to grow and not reach a fully developed stage until other factors, such as fuel or ventilation limits, come into play. See Chapter 7.

Decay

All fires eventually decay and burn out, which represents the final stage of a fire curve. The fire may change from ventilation controlled to fuel controlled at the onset of this stage. This stage always contains the decline of the fire, but is not limited to just the end of a fire. Many fires are dynamic and have the ability to grow and decay multiple times due the amount of fuel remaining, ignition of other fuels, changing ventilation conditions, and initiation of manual or automatic suppression activities.

There are a number of different methods for developing a model of the decay stage. Fire test data usually provide the most accurate way to quantify the decay stage. A constant rate of decay can also be used (SFPE, 2004). Researchers at Factory Mutual (Yu et al., 1994) have developed some correlations for decay rates for rack storage fires. See Chapter 7 for information about predicting the decline in fire size resulting from suppression.

In many performance-based assessments, the decay stage is ignored. The decay stage is usually discounted when it is assumed that all fuel is consumed during the fully developed level. This condition can produce an upper bound of potential exposures. Decay can also be ignored where the conditions of interest (e.g., peak temperatures, smoke production, egress) occur prior to the onset of decay.

Fires Involving Multiple Fuels

In a room, multiple sources of fuel or a larger array of fuel may exist than can be quantified by sources of available data. For example, a wardrobe, chair, and bed may exist in a room. One will be the initial source of the fire, while the others are fuel that is eventually ignited either from direct exposure to the burning item or by the hot upper layer. The heat release rates from the additional burning items are usually treated as additive inputs to the initial burning rate at the time when they are predicted to begin burning.

In these cases, it is necessary to estimate or calculate when the items are ignited. If they are ignited individually by direct exposures, the heat release rates for each fuel item would be added together at the time(s) that the subsequent ignition(s) occurs. If they are all ignited at the same time, the design fire would be the sum of the heat release rates of all items burning.

Consideration must be given to the time at which successive fuels are ignited because of the difference upon the heat release rate curve. If multiple items ignite in rapid succession, a high but narrow curve can be produced, as shown in Figure 4.3. In Figure 4.3, each object individually would have a 1,000 kW peak rate and ignite 40 s apart. The peak rate is 2,500 kW, which is greater than the individual rates of the fuels.

Conversely, if items ignite more slowly, a relatively flat but broad curve can be produced, as shown in Figure 4.4. In Figure 4.4, the objects are the same as those in Figure 4.3, but ignite 120 s apart, rather than 40 s, as in Figure 4.3. The fire burns roughly 50% longer, with a peak rate roughly 50% lower than in Figure 4.3. The resulting peak for the delayed ignition scenario is not significantly higher than for each item burned in isolation

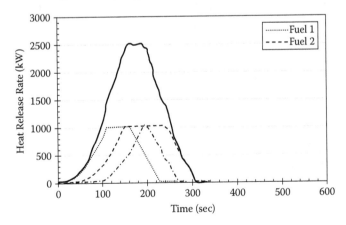

Figure 4.3 High, narrow heat release rate profile characteristic of rapid ignition of additional objects.

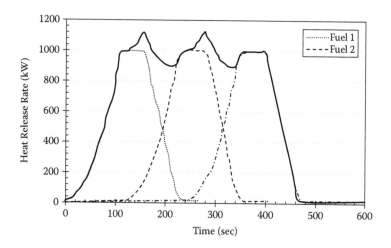

Figure 4.4 Low, broad heat release rate profile characteristic of slow ignition of additional objects.

(1,000 kW). Also, a substantial fire still burns after the first item ignited has extinguished itself. The fuel packages ignite 120 s apart.

FIRE CURVE CREATION METHODOLOGIES

There are a number of ways the fire curve can be produced, including testing (large and small scale), correlations, and analytical approaches. Generally, a combination of these methods is needed to provide information on all four stages of the fire curve. Each method has its advantages and disadvantages. While each method may be used to calculate a fire curve deterministically, they can also be used to select values for risk/probabilistic-type analyses.

Typically, analytical approaches are used to augment data from large- and small-scale tests. Analytical approaches may be used to predict ignition of additional items in a space or the spread of smoldering, if this is an issue. Analytical approaches are mostly used in the growth and fully developed stages of fire. Simplified decay curves are occasionally used.

When the growth stage is calculated analytically, it is typically done with the power law equation:

$$\dot{Q} = \alpha t^p \tag{4.1}$$

where $\dot{Q}$ is heat release rate (kW or BTU/s), α is proportionality constant (kW/s^p or Btu/s^{p+1}), t is time (s), and p is exponent of growth (–).

A fire growing outward radially at a constant rate should have an exponent of $p = 2$ because the area of the involved circular surface grows with the square of the radius, where the radius is predicted to grow linearly in time. Testing has shown that this exponent is valid for a wide range of materials and objects (Heskestad and Delichatsios, 1977a, 1977b, 1979). Compilations of such data can be found in Custer et al. (2008), NFPA 72 (2013), NFPA 204 (2012b), and NFPA 92 (2012a).

These fires are commonly referred to as t-squared fires. Not all fires behave as t-squared fires, however. Fires that grow in three dimensions can have a higher exponent, whereas those that grow in one dimension, such as cable tray fires, can behave linearly ($p = 1$).

The coefficient α in the power law equation denotes the speed at which the fire grows. It is normally considered a constant, but it can change with time. This term is usually determined experimentally through large-scale testing.

This rate of fire growth can also be expressed as a quantity known as the critical time (t_{cr}). The critical time is the time needed for a fire to reach a size of 1,000 BTU/s (1,055 kW). Then, α would be the constant coefficient needed to relate the time and heat release rate together as a curve through the point. For example, for a fire that reaches 1,000 BTU/s (1,055 kW) at a critical time of 150 s:

$$\alpha = \frac{\dot{Q}}{t_{cr}^p} = \frac{1{,}000 \text{ BTU/s}}{[150 \text{ s}]^2} = 0.0444 \text{ BTU/s}^3 \qquad (4.2)$$

Historically, several prototypical growth rates have been named (Alpert, 2008). These names include ultra-fast, fast, medium, and slow, which refer to critical times of 75, 150, 300, and 600 s, respectively. NFPA 72 (2013) adopted modified definitions. NFPA 72 defines these fires as follows: fast fires have a critical time of less than 150 s, medium fires have a critical time between 150 and 400 s, while slow fires have a critical time of 400 s or greater. NFPA no longer uses the designation ultra-fast.

When developing design fires, a bounding t-squared fire is often used to approximate the growth rate provided by test data. Figure 4.5 shows some heat release rate data for a burning sofa. The actual test data during growth can be represented as an idealized t-squared fire to simplify the analysis. The idealized curve tries to match the data as reasonably as possible.

Other times, there may be differences between what was tested and the actual design fire. In this case, the power law approximation is used to bound the available data.

Power law equations are also used when specific knowledge of the fire is limited or unknown. In these situations, the fire scenario represents either situations that have not occurred or widely varying situations. For these

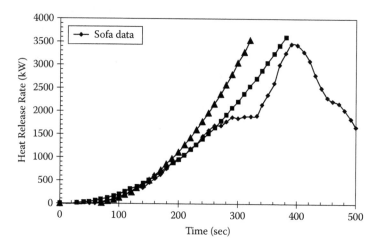

Figure 4.5 Idealized and bounding *t*-squared fire.

situations, a bounding estimate of the fire should be used. Frequently, there is enough knowledge about the situation to make a reasonable bounding estimate. For many forms of fuel, including furniture and storage materials, Tewarson (2008) or NFPA 72 (2013) provide fire growth critical times and maximum heat release rates, as well as other information that will assist in estimates of design fires.

The fully developed stage of a fire can also be calculated analytically. The design fire for the fully developed stage is usually broken into two portions: peak heat release rate and duration of the stage. The peak heat release rate will be either fuel or ventilation limited. The fuel-limited peak heat release rate is usually calculated by multiplying the heat release rate per unit area determined through tests (large or small scale) by the burning area.

If the fire will become ventilation limited, methodologies that can adapt the peak heat release rate to the ventilation conditions are used. This is because the fire can be greater than the ventilation limits during the early portion of a fire due to excess oxygen already present in the room. More detail is available in SFPE (2004). Under steady state conditions, the fire size is based on the ventilation conditions for the room. For most fuels, the fire will produce a heat release rate of approximately 13,100 kJ/kg of oxygen or 3,000 kJ/kg of airflow that is available (Tewarson, 2008). If there is only mechanical ventilation (i.e., no natural ventilation openings such as open doors or windows), the ventilation-limited burning rate is

$$\dot{Q}_{VL,mech} = 3,000 \frac{kJ}{kg\ air} \times \dot{m}_{mech} \tag{4.3}$$

where $\dot{Q}_{VL,mech}$ is mechanical ventilation limited heat release rate [kW], and $\dot{m}_{mech}$ is mechanical ventilation rate (kg air/s).

For natural ventilation, the stoichiometric limit can be calculated as (Walton and Thomas, 2008)

$$\dot{Q}_{VL,nat} = 1,500 \frac{kW}{m^{5/2}} A_o \sqrt{h_o} \qquad (4.4)$$

where $\dot{Q}_{VL,nat}$ is natural ventilation limited heat release rate (kW), A_o is surface area of the opening (m²), and h_o is height of the opening (m).

An approximation for multiple vents is available when the mid-height of the vents is at the same height. If the mid-heights vary significantly, the approximation loses its validity. Multiple vents can be approximated by calculating the total ventilation parameter as the sum of the individual vents:

$$A_o \sqrt{h_o} = \sum_{i=1}^{n} A_{oi} \sqrt{h_{oi}} \qquad (4.5)$$

where n is number of vents (–).

Combined mechanical and natural ventilation can be idealized in the case where the mechanical supply and exhaust in a room are balanced by calculating an equivalent natural ventilation term for mechanical ventilation by dividing the mass flow rate by 0.5 kg/m⁵ᐟ²s:

$$\left(A_o \sqrt{h_o} \right)_{mech} = \frac{\dot{m}_{mech}}{0.5 \, kg \, air/m^{5/2} s} \qquad (4.6)$$

Example

1. What size fire will a room with a 0.8 m wide by 2 m high door support?
2. What size fire will a room with 0.25 m³/s (530 cfm) of supply and exhaust (door closed) support?
3. What size fire will a room support with the door, mechanical supply, and exhaust flowing?

Natural ventilation limit (Equation 4.4):

$$\dot{Q}_{VL,nat} = 1,500 \frac{kW}{m^{5/2}} 1.6\sqrt{2} = 3400 \, kW$$

Therefore, ventilation supports approximately a 3.5 MW fully developed fire.

Mechanical ventilation limit (Equation 4.3):

$$\dot{Q}_{VL,mech} = 3,000 \frac{kJ}{kg\ air} \times 0.25 \frac{m^3}{s} \times 1.2 \frac{kg\ air}{m^3} = 900\ kW$$

Therefore, ventilation supports approximately a 1 MW fully developed fire.

Equivalent natural ventilation limit (Equation 4.6):

$$\left(A_o\sqrt{h_o}\right)_{mech} = \frac{0.25\frac{m^3}{s} \times 1.2\frac{kg\ air}{m^3}}{0.5\,kg\ air/m^{5/2}s} = 0.6m^{5/2}$$

Combine natural and mechanical ventilation (Equation 4.5):

$$\sum_{i=1}^{2} A_{o_i}\sqrt{h_{o_i}} = 1.6\sqrt{2} + 0.6 = 2.86\ m^{5/2}.$$

Determine ventilation limit (Equation 4.4):

$$\dot{Q}_{VL,nat} = 1500 \times 2.86 = 4,300\ kW$$

This value can also be calculated as the sum of the independent natural ventilation-limited and mechanical ventilation-limited heat release rates (i.e., 3400 kW + 900 kW).

The duration of the fully developed region can be calculated analytically. Occasionally, there are enough large-scale data or readily available tabulated quantities to calculate the duration directly. For example, the duration of a pool fire can be calculated by dividing the liquid depth by the regression rate, when these quantities are known.

Example

How long will a 1 m diameter gasoline fire, 5 cm in depth, burn?
From Gottuk and White (2008), the regression rate for a 1 m diameter gasoline pool is 5 mm/min. Therefore

$$t_{burn} = 5\ cm/5\ mm/min = 10\ min$$

Typically, the duration is calculated indirectly based on the amount of mass consumed during this stage of the fire. This requires that the mass consumed from other stages of the fire be calculated first and subtracted from the total consumable mass. The duration is calculated by multiplying the amount of mass that is consumed during this stage by the heat of combustion and dividing by the peak heat release rate calculated above.

Example

How long will the idealized sofa in Figure 4.5 burn if the ventilation limit is 1.5 MW? Assume the couch weighs 70 kg with an average heat of combustion of 17 kJ/g and a linear decay of 17 kW/s.

The growth phase lasts until the ventilation limit is reached. This is calculated by solving for time in Equation 4.1. Using the bounding curve in Figure 4.5, this time is

$$t_{growth} = \sqrt{\frac{\dot{Q}}{\frac{1055}{t_{cr}^2}}} = \sqrt{\frac{1500}{\frac{1055}{200^2}}} = 238 \, sec$$

Integrating the idealized t-squared curve for heat release rate yields the total energy consumed during the growth period

$$E = \frac{1055}{3t_{cr}^2} t_{growth}^3 = \frac{1055}{3(200)^2} 238^3 = 119,000 \, kJ$$

Dividing this by the heat of combustion yields the total mass consumed during the growth phase:

$$m_{total} = \frac{E}{H_c} = \frac{119000}{17} = 7 kg$$

This yields a mass loss of 7 kg during the growth phase. The decay portion is calculated essentially the same way. Decay lasts 88 s:

$$t_{decay} = \frac{\dot{Q}}{\dot{Q}_{decay}} = \frac{1500}{17} = 88 \, sec$$

Integrating the linear decay yields the total energy consumed:

$$E = \frac{\dot{Q} t_{decay}}{2} = \frac{1500 \times 88}{2} = 66,000 \, kJ$$

Dividing this by the heat of combustion yields the total mass consumed during the decay phase:

$$m_{total} = \frac{E}{H_c} = \frac{66000}{17} = 3.9 \, kg$$

Therefore, a mass of 3.9 kg is consumed during the decay phase and the total mass loss during growth and decay is 10.9 kg. This leaves

59.1 kg of mass to be consumed during the fully developed stage. The duration of the fully developed phase is

$$t = \frac{mH_c}{\dot{Q}} = \frac{59.1(17000)}{1500} = 670 \text{ sec}$$

See Figure 4.6 for depiction.

The decay stage can also be calculated analytically. Occasionally, large-scale data will be used to determine the decay stage. If suppression is present, correlations can be used to calculate the decay. In lieu of any other known type of decay, either linear or no decay is assumed.

Data from small-scale testing can be used as inputs for analytically determining fire curve characteristics. For example, a heat release rate per unit area obtained from bench-scale tests can be used to determine the fully developed heat release rate of a fire.

Small-scale testing is testing done at a bench-scale size. It includes apparatus such as cone calorimeters, the lateral ignition and flame spread test (LIFT) apparatus, open- and closed-cup flash point testers, and the U.S. Bureau of Mines flammability apparatus. The advantage of bench-scale testing is that it is a relatively easy and cheap method to test a variety of different materials. It is usually not possible to use these data directly in the creation of a fire curve due to the small scale of the testing because fires can have characteristics that change with size. Rather, the results are typically used in analytical approaches.

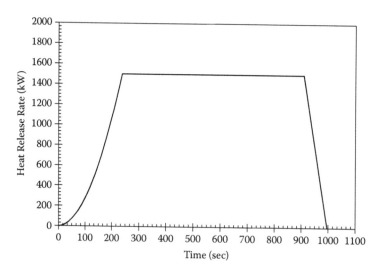

Figure 4.6 Design fire curve for a sofa fire.

Small-scale testing is able to provide ignition temperatures and critical heat fluxes for solids, ignition temperatures for liquids, and flammable concentration limits of gases. Due to the small size, information on the growth stage of a fire cannot be provided. Tests at this scale are capable of providing heat release rate per unit area, which can be used to calculate peak heat release rates in the fully developed stage. Combustion product yields are also possible with small-scale testing.

Example

Determine the fire curve for two 1 m cubes made of 12.7 mm (1/2 in.) plywood that are located in a large warehouse, 3 m apart. Each cube has a mass of 42 kg.

First, a fire curve for a single plywood box (cube) is developed. It is assumed that ignition occurs for the first box. No suitable large-scale test data for plywood boxes could be located, so an estimation of the growth rate needs to be made. There are some large-scale test data on stacks of wood pallets (NFPA, 2013), which are about the same size as the plywood boxes. The wood pallet tests are expected to bound the plywood boxes, but not be overly conservative. The test data on the wood pallets indicate that they can be idealized as t-squared fires with a critical time for 0.5 m (1.5 ft) high stacks between 150 and 310 s, while 1.5 m (5 ft) high stacks are between 90 and 190 s. The 1 m high box is about halfway in between these two pallet stacks in terms of height. An average of the average critical time for the two heights was calculated (185 s) to give a bounding estimate of the growth time of the box. This corresponds to a fire growth coefficient (α) of 0.031 kW/s^2.

The large-scale test data do not contain any information on peak heat release rates to define the fully developed portion of the fire. Suitable small-scale test data from the cone calorimeter were able to be located for plywood (Tran, 1991). The tests determined that the 1 min average heat release rate at 40 kW/m^2 irradiance is 150 kW/m^2 for vertically oriented plywood and 195 kW/m^2 for horizontally oriented plywood. The boxes contain 4 m^2 of vertically oriented plywood and 1 m^2 of horizontally oriented plywood. This configuration gives a peak heat release rate of 795 kW (150 × 4 + 195 × 1). The bottom of the box is assumed not to burn.

Fire duration for the plywood cube was estimated using the same methodology as the previous sofa fire duration example. An effective heat of combustion of 13 kJ/g was obtained from the cone calorimeter data. From this methodology, approximately 3.2 kg is consumed in the 160 s growth phase, and the remaining 31.8 kg is consumed in 520 s (assuming the bottom of the box does not burn and no decay). See Figure 4.7 for depiction of fire curve.

With the first fire curve developed, a determination of fire spread to the adjacent box needs to be determined. The method in SFPE (1999) was used to determine the radiative flux the second object experiences

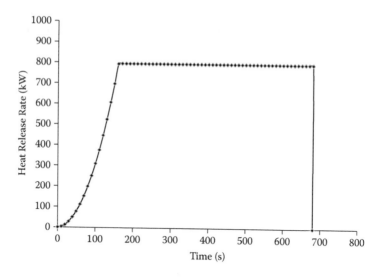

Figure 4.7 Fire curve for plywood box example.

from the first fire. The box itself is idealized as a pool fire with an area of 1 m² (same plan area as the box). This yields a flame height of 2.17 m. The flame was idealized as a circular cylinder in order to calculate the radiation exchange from the flame to the half height of the flame (approximately the top of the box). That calculation determined the radiation shape factor was 0.064. Using the Shokri and Beyler (Beyler, 2008a) flame heat flux correlation with the recommended safety factor of 2 yields a heat flux to the box of 7.3 kW/m². From Tewarson (2008), the critical heat flux is 10 kW/m². Propagation from one box to the other is not expected. The fire would consume just one box.

Large-scale tests can be used directly to estimate fire curves. If there are test data that directly correlate to a chosen fire scenario, using them as input is appropriate. For example, the burning of a dry Christmas tree could be based on data in Ahonen et al. (1984). Using test data directly requires detailed knowledge on the fuel package.

Alternative large-scale tests can be used as a guide to determine the characteristics of the various elements of the design fire. In most cases, large-scale tests are provided to estimate the growth rate or intensity of the fully developed stage. For example, the peak heat release rate of a trash bag fire can be estimated based on data provided in Lee (1985). A fire growth rate can be estimated from the large-scale test data summarized in NFPA 72 (2013).

Large-scale testing is testing done at full scale or close to full scale. It is the best way to determine a fire curve since actual samples of the material are used experimentally in a representative geometric configuration. Conversely, this specificity is also the biggest disadvantage of large-scale testing because it can limit applicability outside the test configuration. Large-scale tests are also expensive to set up, instrument, and run, which potentially limits the number of trials that can be performed. Theoretically, these tests are able to produce data on all four stages of the fire curve. Realistically, most large-scale tests focus on one or more stages of the fire, but do not always provide the instrumentation to measure data at every stage. For instance, ignition is often performed with a convenient laboratory device, which may differ from potential ignition methods in the field. In addition to the four stages, large-scale testing can also provide data for combustion product yields for some tests.

An example where large-scale data might be used is a tenability analysis for an apartment complex. The objective would be to determine if there is enough time for everyone to exit the building before untenable conditions are reached (e.g., heat, smoke, carbon monoxide). One fire scenario that would need to be evaluated is a living room fire involving a couch.

Large-scale data can be used to determine the fire curve for this couch. Often the data are idealized to simplify data entry. Figure 4.8 shows the measured test data along with the idealized curve developed. This curve could then be entered into a fire model that would be able to calculate tenability conditions throughout this and other apartments.

In addition to the heat release rate data, the tenability analysis would need yield data to determine tenability conditions. In this case, yield data

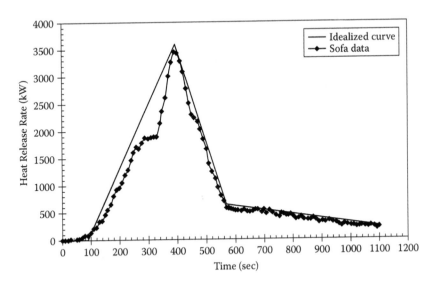

Figure 4.8 Fire curve for sofa fire. (From Madrzykowski, D., and Vettori, R., A Sprinkler Fire Suppression Algorithm for the GSA Engineering Fire Assessment System, NISTIR 4833, National Institute of Standards and Technology, Gaithersburg, MD, 1992.)

(e.g., smoke particulate or carbon monoxide production) were not provided with the heat release rate curve; these data would need to be obtained through other means, such as cone calorimeter data.

Correlations can also be used to determine the various stages of a fire curve. Correlations are usually based on test data, although some based on fire modeling also exist. Some correlations can only be used to create limited portions of the fire curve, while others, like the NIST furniture correlations (Krasny et al., 2001), can produce an entire fire curve.

Other correlations are used to calculate only one or two stages of the fire curve. Examples of these correlations (Babrauskas, 2008) include peak burning rate correlations for wood cribs and wood pallets, and sprinkler suppression correlations, which calculate the decay of a fire due to suppression.

Like all approaches, each correlation contains its own assumptions and limitations. These correlations should only be used for situations that meet these limitations and assumptions.

Example

Determine the fire curve for a 1.5 m (5 ft) high stack of pallets.

The growth rate of a 1.5 m (5 ft) high stack of pallets can be determined from large-scale test data (NFPA, 2013). The critical time ranges between 90 and 190 s and depends on the type of analysis being performed (e.g., time to hazard). To be conservative, the shortest critical time (90 s) is chosen.

The peak heat release rate is determined based on a correlation (Babrauskas, 2008). Assuming a moisture content of 9%, the peak heat release rate is calculated as follows:

$$Q = 1450(1 + 2.14H)(1 - 0.027M)$$

$$= 1450(1 + 2.14(1.5))(1 - 0.027(9))$$

$$= 4,600 \text{ kW}$$

Assuming pallets weigh 110 kg/m of height, the duration of the fire is 435 s (188 for growth, 247 for fully developed) assuming no decay. See Figure 4.9 for a depiction of the estimated heat release rate.

Uncertainty exists in every parameter and methodology used to calculate the fire curve. There are a number of methods for dealing with these uncertainties. These methods include the use of safety factors, classical uncertainty analysis, margin analysis, importance analysis, sensitivity analysis, parametric analysis, and comparative analysis. The use of these tools may require modifications or refinements to the fire curve. In some cases, uncertainty can ease calculation of a fire curve since an apparent conservative

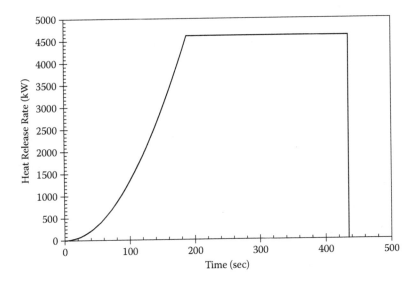

Figure 4.9 Fire curve for the stack of pallets example.

fire is easier to estimate than an exact one. However, it must be confirmed that the predicted fire is conservative. See Chapter 12.

IMPACT OF ANALYSIS METHODOLOGY SELECTION

The method utilized to evaluate the effect of design fires will impact the required information for each design fire scenario. Understanding the inputs to the calculations and what is known about the problem is crucial. Different methodologies require different inputs. It is wise to choose methodologies that require the same inputs as what is known about the problem. Conversely, if certain parameters are not known very well, methodologies that do not use these parameters or are relatively insensitive to those parameters should be selected.

Once a methodology is selected, the design fire curve can be constructed. The choice of method determines which stages of the fire curve actually can be established. Like the selection of trial designs, which can be an iterative process, the construction of the fire curve is iterative as well. This is due to discoveries about enclosure effects such as ventilation and radiative feedback as calculations begin, as well as uncertainty in the calculation and methodology input parameters. It may not be possible to predict the time or effect that these parameters have until the fire curve is actually used to evaluate the trial designs.

REFERENCES

Ahonen, A., Kokkala, M., and Weckman, H., *Burning Characteristics of Potential Ignition Sources of Room Fires*, Research Report 285, Technical Research Centre of Finland, Espoo, 1984.

Alpert, R., Ceiling Jet Flows, in *SFPE Handbook of Fire Protection Engineering*, 4th ed., National Fire Protection Association, Quincy, MA, 2008.

Babrauskas, V., Heat Release Rates, in *SFPE Handbook of Fire Protection Engineering*, 4th ed., National Fire Protection Association, Quincy, MA, 2008.

Beyler, C., Fire Hazard Calculation for Large Open Hydrocarbon Fires, in *SFPE Handbook of Fire Protection Engineering*, 4th ed., National Fire Protection Association, Quincy, MA, 2008a.

Beyler, C., Flammability Limits of Premixed and Diffusion Flames, in *SFPE Handbook of Fire Protection Engineering*, 4th ed., National Fire Protection Association, Quincy, MA, 2008b.

Custer, R., Shifiliti, R., and Meacham, B., Design of Detection Systems, in *SFPE Handbook of Fire Protection Engineering*, 4th ed., National Fire Protection Association, Quincy, MA, 2008.

Drysdale, D., *An Introduction to Fire Dynamics*, 3rd ed., John Wiley, New York, 2011.

Gottuk, D., and White, D., Liquid Fuel Fires, in *SFPE Handbook of Fire Protection Engineering*, 4th ed., National Fire Protection Association, Quincy, MA, 2008.

Hasemi, Y., Surface Flame Spread, in *SFPE Handbook of Fire Protection Engineering*, 4th ed., National Fire Protection Association, Quincy, MA, 2008.

Heskestad, G., and Delichatsios, M., *Environments of Fire Detectors—Phase 1: Effect of Fire Size, Ceiling Height and Material: Measurements*, vol. 1, NBS-GCR-77-86, National Bureau of Standards, Washington, DC, 1977a.

Heskestad, G., and Delichatsios, M., *Environments of Fire Detectors—Phase 1: Effect of Fire Size, Ceiling Height and Material: Analysis*, vol. 2, NBS-GCR-77-95, National Bureau of Standards, Washington, DC, 1977b.

Heskestad, G., and Delichatsios, M., The Initial Convective Flow in Fire, presented at Seventeenth Symposium (International) on Combustion, The Combustion Institute, Pittsburgh, PA, 1979.

Hurley, M.J., and Rosenbaum, C.R. Performance-Based Design, in *SFPE Handbook of Fire Protection Engineering*, 4th ed., National Fire Protection Association, Quincy, MA, 2008.

Krasny, J., Parker, W., and Babrauskas, V., *Fire Behavior of Upholstered Furniture and Mattresses*, William Andrew, Norwich, NY, 2001.

Lattimer, B., Heat Fluxes from Fires to Surfaces, in *SFPE Handbook of Fire Protection Engineering*, 4th ed., National Fire Protection Association, Quincy, MA, 2008.

Lee, B., *Heat Release Rate Characteristics of Some Combustible Fuel Sources in Nuclear Power Plants*, NBSIR 85-3195, National Bureau of Standards, Gaithersburg, MD, 1985.

Madrzykowski, D., and Vettori, R., *A Sprinkler Fire Suppression Algorithm for the GSA Engineering Fire Assessment System*, NISTIR 4833, National Institute of Standards and Technology, Gaithersburg, MD, 1992.

NFPA, *Standard for Smoke Control Systems*, NFPA 92, National Fire Protection Association, Quincy, MA, 2012a.

NFPA, *Standard for Smoke and Heat Venting*, NFPA 204, National Fire Protection Association, Quincy, MA, 2012b.

NFPA, *National Fire Alarm Code*, NFPA 72, National Fire Protection Association, Quincy, MA, 2013.

SFPE, *Engineering Guide Assessing Flame Radiation to External Targets from Pool Fires*, Society of Fire Protection Engineers, Bethesda, MD, 1999.

SFPE, *SFPE Engineering Guide to Fire Exposures to Structural Elements*, Society of Fire Protection Engineers, Bethesda, MD, 2004.

Tewarson, A., Generation of Heat and Chemical Compounds in Fires, *in SFPE Handbook of Fire Protection Engineering*, 4th ed., National Fire Protection Association, Quincy, MA, 2008.

Torero, J., Flaming Ignition of Solid Fuels, in *SFPE Handbook of Fire Protection Engineering*, 4th ed., National Fire Protection Association, Quincy, MA, 2008.

Tran, H., Experimental Data on Wood Materials, in *Heat Release in Fires*, Elsevier Applied Science, New York, 1991.

Walton, W., and Thomas, P., Estimating Temperatures in Compartment Fires, in *SFPE Handbook of Fire Protection Engineering*, 4th ed., National Fire Protection Association, Quincy, MA, 2008.

Yu, H., Lee, J., and Kung, H.C., Suppression of Rack-Storage Fires by Water, presented at Fire Safety Science—Proceedings of the Fourth International Symposium, International Association for Fire Safety Science, London, 1994.

Chapter 5

Fire Dynamics and
Hazard Calculations

INTRODUCTION

Engineering tools, such as models, are typically used to perform two types of tasks: (1) model a fire and its effects in a design fire scenario and (2) evaluate a trial design strategy to determine if it achieves project goals, objectives, and performance criteria. Modeling a fire and its effects in a design basis fire scenario is part of a fire hazards analysis. The purpose of conducting a fire hazards analysis is to determine the conditions that could occur in the event of a fire: temperature, heat flux, concentration of products of combustion, etc. These conditions can be determined by using fire models.

These same types of tools would also be used to evaluate a trial design strategy. Evaluations determine if a design provides an adequate level of safety. Additionally, the effects of trial fire protection strategies would be considered in an evaluation.

A key element of performance-based design is estimation of the hazards that result from a fire and their effects on targets, where targets are the items that are intended to be protected by the design. Examples of targets include people and the items that are stored within a building.

This chapter provides an overview of the types of hazards calculations that may be performed and the effects that these hazards may have on targets other than people. The effects of hazards on people are discussed in detail in Chapter 6. The following hazards may be of interest:

- Fire environment temperature
- Heat flux
- Smoke production
- Fire plume and ceiling jet temperatures and velocities
- Species production
- Depth of upper layer

There are a variety of methods that can be used to predict hazards and their effects. A comprehensive discussion of how hazards and their effects can be determined is beyond the scope of this chapter (and indeed, this book). Therefore, this chapter will provide a simple hazards assessment methodology and identify some other potential hazards assessment methods.

Typically, the prediction of hazards involves determining conditions that occur under nonsteady conditions. Smoke filling in an enclosed space is an example. Smoke filling of a space is like filling a tub, although smoke fills a room from the top instead of from the bottom. In this case, the volumetric flow rate that fills the tub is independent of the water level. Stated mathematically:

$$\frac{dL}{dt} = \frac{\dot{V}}{A}$$

where L is water level, $\dot{V}$ is volumetric flow rate of water, and A is cross-sectional area of the tub.

If the problem is simplified by assuming that the tub does not vary in cross-sectional area with height, then this equation can be integrated easily, since the volumetric flow rate of water into the tub is constant.

The same equation could be used to determine the rate at which the elevation of a smoke layer changes in a room that is filling with smoke. The rate of smoke filling of an atrium can be calculated using the following equation (Budnick et al., 2008):

$$\frac{dL}{dt} = \frac{1}{\rho A}\left[0.065\dot{Q}_c^{\frac{1}{3}}Z^{\frac{5}{3}}\right]$$

where L is level of smoke layer interface, ρ is density of air, $\dot{Q}_c$ is convective portion of heat release rate, and Z is elevation of smoke layer interface above the fire.

This equation cannot be easily integrated because the rate of change of the smoke level interface depends on conditions that change during the course of smoke filling. The equation can be integrated if one can make some simplifying assumptions. Well-characterized fires, such as fires that grow according to a power law curve, provide just such a simplification. However, in other cases, including fires that are not well characterized, the equation must be solved numerically. A potential numerical method for these cases involves use of a Euler solution technique (discussed in more detail at the end of this chapter).

Many fire problems cannot be solved with closed-form solutions, so other solution techniques must be used. Computer models automate fire dynamics calculations. In some cases, computer models solve simple algebraic correlations that could be solved by hand (e.g., NIST, 1995). In others, computer models are used to numerically solve differential equations for which closed-form solutions do not exist.

A large number of computer fire models are available and have been documented by a survey (Olenick and Carpenter, 2003). The modeler must determine which ones are appropriate to a given situation and which are not. The key to this decision is a thorough understanding of the assumptions and limitations of the individual model or calculation and how these relate to the situation being analyzed. Beyler and DiNenno (2008) provide a discussion of the types of computer models that are available and their limitations.

In some cases, models may not be sufficient for a given application, and one must resort to testing. Testing is discussed in more detail in Chapter 10.

FIRE ENVIRONMENT TEMPERATURE

The fire environment temperature may be of interest in cases where tenability is involved or where it is desired to maintain temperatures below a level that would damage protected items.

If a fire is in the open or in a very large space (such as a sports arena), then the fire may be treated as a plume, and the effects associated with an enclosure neglected.

Example

A steel beam is located 6 m above floor level in a very large atrium. If performance criteria for the beam requires that it not reach a temperature of 550°C, could the performance criteria be exceeded if the design fire is 5 MW?

From Budnick et al. (2008),

$$\Delta T = 18.8 \dot{Q}^{2/3} (h - z_0)^{-5/3}$$

If virtual origin is neglected ($z_0 = 0$), assuming that a smoke layer will not form and assuming an ambient temperature of 20°C, the temperature 6 m above floor level would be 277°C. Therefore, it would not be possible for the beam to be heated above 277°C, regardless of the length of exposure, and the performance criteria would be achieved.

Example

A room measures 3 m by 4 m and has a ceiling height of 2.5 m. There is a single door into the room that is 1 m wide and 2 m in height. The room is lined with 13 mm thick gypsum board. If there is a 1 MW steady fire in the room, what would be the temperature in the room 1 min after ignition?

The first step would be to determine whether flashover might occur in the room. From Walton and Thomas (2008):

$$\dot{Q} = 750 A_o \sqrt{H_o}$$

where $\dot{Q}$ is heat release rate necessary to cause flashover (kW), A_o is area of ventilation opening (m²), and H_o is height of ventilation opening (m).

The heat release rate necessary to cause flashover is 2,100 kW.

Budnick et al. (2008) provide a method for calculating pre-flashover compartment temperatures. This method predicts the average upper layer temperature (assuming that a hot, upper layer forms and the lower layer remains at ambient temperature).

From Budnick et al. (2008), the thermal penetration time is

$$t_p = \left(\frac{\rho c}{k} \right) \left(\frac{\delta}{2} \right)^2$$

where ρ is density of enclosure lining (kg/m³), c is specific heat of enclosure lining (kJ/kg °C), k is thermal conductivity of enclosure lining (kW/m °C), and δ is thickness of enclosure lining (m).

For this example, the thermal penetration time is 248 s (the thermal penetration time is a measure of whether the conduction through the lining has reached steady state). Therefore, since the time of interest (1 min) is less than the thermal penetration time, the enclosure conductance can be calculated as follows:

$$h_k = \left(\frac{k\rho c}{t} \right)^{\frac{1}{2}} \quad \text{or} \quad h_k = 0.057 \text{ kW/(m²K)}$$

The temperature rise is then calculated as

$$\Delta T = 6.85 \left(\frac{\dot{Q}^2}{h_k A_s A_o \sqrt{H_o}} \right)^{\frac{1}{3}}$$

where $\dot{Q}$ is heat release rate (kW), and A_s is surface area of enclosure (m²).

The temperature rise would be 323°C. Note that it was only possible to determine the temperature rise in this room because a steady fire was assumed. If the fire were not steady, a computer fire model would have to be used to determine the temperature in the compartment.

HEAT FLUX

In fires, it is not the temperature of the surrounding gases that always presents a hazard, but rather the heat flux to the target that can cause damage. If an object is not immersed in hot gases, the primary mode of heat transfer would be from radiation. This would be the case for an item that is outside of a fire plume and is not submerged in a hot smoke layer. If the object is immersed in hot gases, heat transfer by convection would play a role as well.

For an item that is adjacent to a burning object but is not immersed in hot gases, there are a number of methods available for calculating the radiative heat flux exposure (Budnick et al., 2008). One such case would involve an object located below a hot upper layer heated solely by radiation. The total intensity of radiation emitted by a burning object can be calculated as follows:

$$\dot{q}_r'' = \frac{\chi_r \dot{Q}}{4\pi R_0^2}$$

where χ_r is radiative fraction (–), $\dot{Q}$ is heat release rate (kW), and R_0 is distance from center of the base of the fire to the target (m).

SMOKE PRODUCTION

Smoke is primarily constituted of air—the concentration of combustion products, such as CO_2, H_2O, CO, etc., is much less than the amount of air that is entrained into the fire plume (Budnick et al., 2008). Therefore, the properties of combustion products are typically neglected when determining the production rate of smoke. The smoke production rate can then be calculated by determining the amount of air entrained into the smoke plume.

Budnick et al. (2008) provide the following equation to calculate the mass entrainment rate into a smoke plume:

$$\dot{m}_s = 0.065 \dot{Q}^{\frac{1}{3}} Y^{\frac{5}{3}}$$

where $\dot{Q}$ is heat release rate (kW), and Y is elevation above the base of the fire (m).

Example

What is the smoke production rate from a 5 MW fire in a 12 m high atrium?

For this example, the mass entrainment rate would be 70 kg/s. This can be by divided by the density of air (1.2 kg/m³) to yield a volumetric mass entrainment rate of 58 m³/s. Note that as the atrium fills with smoke, the smoke layer would descend, and with it, the height over which the plume entrained air. Therefore, the smoke filling rate in this example would decrease with time as the layer descends.

Computer models are available that can be used to calculate the smoke filling rate as a function of time.

SPECIES PRODUCTION

While the production rates of products of combustion are not considered when estimating the production rate of smoke, the products of combustion are what typically make smoke hazardous. The products of combustion can vary depending upon the fuel that is burning and the ventilation available to the fuel. Species production can be estimated using the following equation (Hurley and Bukowski, 2008):

$$\dot{G}_i = y_i \frac{\dot{Q}}{\Delta H_c}$$

where $\dot{G}_i$ is production rate of species i (kg/s), y_i is yield fraction of species j (–), $\dot{Q}$ is heat release rate (kW), and ΔH_c is heat of combustion of fuel (kJ/kg).

Where ventilation conditions vary over the course of a fire (e.g., the fire transitions from fuel controlled to ventilation controlled), the species production rates can change.

Example

A sofa constructed of GM21 flexible polyurethane foam burns with a heat release rate of 1 MW. If the fire is well ventilated, what is the production rate of CO?

From Tewarson (2008), the effective heat of combustion of GM21 is 17,800 kJ/kg, and the CO yield is 0.010 g/g. Therefore, the production rate of CO would be 0.6 mg/s.

In order to estimate the species concentrations resulting from a fire, the species production rate can be integrated with respect to time to obtain the total resulting mass. This mass can then be divided by the total mass of the smoke layer (obtained from air entrainment) to obtain the concentration.

In order to quantify the effect of changing ventilation conditions, it would be necessary to consider the equivalence ratio of the fire environment. Gottuk and Lattimer (2008) provide an overview of this approach.

SELECTING A MODELING APPROACH

With many modeling techniques available, selecting one for a given application can be a daunting task. While the tools available vary in sophistication,

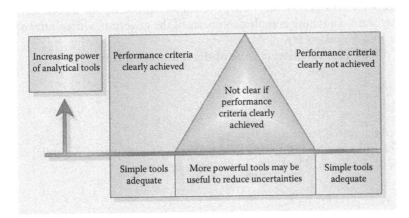

Figure 5.1 Idealized framework for choosing a model. (From Mowrer, F., The Right Tool for the Job, *Fire Protection Engineering*, 13, 2002. Used with permission of Society of Fire Protection Engineers, copyright © 2002.)

a more sophisticated model is not necessarily a better choice for a given application than a less sophisticated model.

Mowrer (2002) presented the idealized framework for choosing a model shown in Figure 5.1. Mowrer suggests that if a simple model clearly shows that a set of performance criteria are clearly achieved (or not achieved), then the simple model is sufficient. However, if the results are not clear when using the simple model, then one would need to choose a more complex model.

This is an important distinction, because more sophisticated models generally take more time and effort to run than simpler models. For example, a simple method may be sufficient to size a simple smoke control system if the heat release rate can be defined using a power law function and the geometry in the atrium is regularly shaped.

If the atrium is irregularly shaped, a simple method may still be appropriate if the difference between the predicted hazards and the performance criteria is large. In other cases, more complex methods, such as zone models or computational fluid dynamics (CFD) models, may be required. However, even when computer models are used, hand calculations could still be used to benchmark the results that were obtained from the computer model. In many cases, using multiple models to evaluate the same scenario and comparing results provides benefits in the evaluation and added confidence to the analysis.

SIMPLE APPROACH FOR NONSTEADY SITUATIONS

Simple computer programs and spreadsheets can be used to perform basic fire hazard calculations. In the case of the equations in this chapter, this is relatively straightforward. However, many fires and fire effects are not steady state, and more complex solutions of the governing differential equations may be necessary.

A differential equation may be of the following form:

$$\frac{dy}{dt} = f(y,t)$$

where the initial value $y(t = 0)$ is y_0.

The Euler method is a numerical technique for solving differential equations of this form. The Euler method can be stated as

$$y_{n+1} = y_n + hf(y_n, t_n)$$

where y_n is value of quantity at time step n, y_{n+1} is value of quantity at time step $n + 1$, and h is time step size.

This process can be iterated over the desired length of time to obtain the desired solution. Since the Euler method determines the value of equation y at time step $n + 1$ based on the value at time step n and the slope of the tangent to y at time step n, errors can be introduced from nonlinearities of equation y. There are methods available to reduce this error, such as the improved Euler method.

Another method that can be used to reduce the error is to decrease the size of the time step, recognizing that as the time step approaches zero, the difference between the predicted value of y and the actual value also approaches zero. The computational power offered by modern computers allows very small time steps to be used to get a solution rapidly, even when very small time steps are used.

The default for many spreadsheets is not to allow iterative calculations. Spreadsheets for which this is the case would need to be configured to allow iteration. The spreadsheet's user manual or help function can be consulted for assistance.

Example

Consider the equation that was provided at the beginning of this chapter for filling a bathtub.

$$\frac{dL}{dt} = \frac{\dot{V}}{A}$$

A Euler solution for this equation would be

$$L_{n+1} = L_n + \Delta T \frac{\dot{V}_n}{A_n}$$

REFERENCES

Babrauskas, V., Heat Release Rates, in *SFPE Handbook of Fire Protection Engineering*, National Fire Protection Association, Quincy, MA, 2008.

Beyler, C., Dinenno, P., Carpenter, D., and Watts, J. Introduction to Fire Modelling, in *Fire Protection Handbook,* National Fire Protection Association, Quincy, MA, 2008.

Budnick, E., et al., Closed Form Enclosure Fire Calculations, in *Fire Protection Handbook*, 20th ed., National Fire Protection Association, Quincy, MA, 2008.

Deal, S., *Technical Reference Guide for FPETool Version 3.2*, NISTIR 5486-1, National Institute of Standards and Technology, Gaithersburg, MD, 1995.

Gottuk, D., and Lattimer, B., Effect of Combustion Conditions on Species Production, in *SFPE Handbook of Fire Protection Engineering*, National Fire Protection Association, Quincy, MA, 2008.

Hurley, M., and Bukowski, R., Fire Hazard Analysis Techniques, in *Fire Protection Handbook*, 20th ed., National Fire Protection Association, Quincy, MA, 2008.

Mowrer, F., The Right Tool for the Job, *Fire Protection Engineering*, 13, 2002.

Olenick, Stephen M., and Carpenter, Douglas J., An Updated International Survey of Computer Models for Fire and Smoke, *Journal of Fire Protection Engineering*, 13(2), 87–110, 2003.

Tewarson, A., Generation of Heat and Gaseous, Liquid and Solid Products in Fires, in *SFPE Handbook of Fire Protection Engineering*, National Fire Protection Association, Quincy, MA, 2008.

Walton, D., and Thomas, P., Estimating Temperatures in Compartment Fires, in *SFPE Handbook of Fire Protection Engineering*, National Fire Protection Association, Quincy, MA, 2008.

Chapter 6

Human Behavior

INTRODUCTION

A few decades ago, the responses of people to fire were not integrated into fire protection design. When human behavior was considered, it was often limited to simple assumptions that were not always appropriate. For example, designers may have assumed that occupants of a building immediately began exiting upon hearing an alarm signal. However, information in the literature indicates that this frequently is not the case (Canter, 1985; Bryan, 1983; Wood, 1980).

Information relative to how people perform in fire first began to emerge in the early 1900s and began to expand significantly in the 1980s and 1990s (Bryan, 1998). But, unlike other areas of fire protection engineering, many aspects of human behavior often can't be reduced to simple equations that can be applied in the general case.

The lack of readily applied correlations contributed to the neglect of human behavior in engineered fire protection design. However, information exists that can be applied and used to improve fire protection designs.

Generally, analysis of how people are impacted by fire is a measure of how the required safe egress time (RSET) compares to the available safe egress time (ASET). For a typical design to be considered acceptable (from a life safety standpoint), the available safe egress time must be at least equal to the required safe egress time:

$ASET \geq RSET$

In calculations involving how people behave in fires, uncertainties can be large. To account for these uncertainties, a factor of safety is frequently applied:

$ASET \geq F \times RSET$

The required safe egress time includes several components:

- The time that it takes for people to become aware of a fire (notification time)
- The time that it takes for people to decide to evacuate (decision time)
- The time that it takes for people to move through the egress system (movement time)

The available safe egress time is a function of the hazards created by the fire environment and the impact of the fire environment on people. It takes time for cues from a fire to become perceptible to building occupants. For example, if a fire is developing in a high-rise building several floors below where occupants are located, it will take time for either smoke to travel to their location or for a fire alarm system to be activated. Additionally, it will take time for the fire to develop to a state where it creates an environment that is hazardous (either immediately or cumulatively) to building occupants. The hazards created by the fire environment (temperature, heat flux, toxic species concentration) are a function of the behavior of the fire. Therefore, the available safe egress time is comprised of two primary components:

- The time that it takes for the fire to create a hazardous environment
- The time that it takes for the fire environment to incapacitate occupants

The former is a function of the fire environment. The latter is a function of the tenability of the hazards created by the fire and how they impact building occupants.

This chapter follows the methodology for considering human behavior in fires that is identified in the *SFPE Engineering Guide—Human Behavior in Fire* (2003). Additionally, another area relevant to considering how people are impacted by fire—tenability analysis—is also addressed.

Aspects of Human Behavior in Fire

The *SFPE Engineering Guide—Human Behavior in Fire* divides analysis of human behavior in fire into four categories:

Occupant characteristics. Occupant characteristics are the features of building occupants that could influence how they respond in the event of a fire. These factors include physical capabilities, sensory capabilities, familiarity with the building, social roles, and commitment to activities. As such, occupant characteristics are the inputs to human behavior analysis.

Response to cues. Analysis of the notification of occupants involves consideration of the types of fire cues to which people are exposed and how people respond to the cue(s). Fire cues include direct indications of a fire (sight or smell of smoke) and indirect indications of fire, such as fire alarm signals, cues from other occupants, and disruptions of building services. The time that it takes for an occupant to become aware of a fire includes the time necessary to receive the cue, the time necessary to recognize the cue, and the time necessary to interpret the cue.

Decision. Once an occupant recognizes a cue as such, he or she will make a decision as to what action to take. The most common decisions include seeking additional information (which is particularly true when the cue does not provide a strong indication that the fire is hazardous), searching for or notifying others (particularly when the occupant is part of a group, such as a family), continuing with the activities in which he or she was previously engaged, or beginning to evacuate.

Movement. Once an occupant has made a decision to evacuate, it will take time for him or her to move to a safe place outside the building or to a protected place within the building. The time that it takes for people to evacuate is a function of the number of people within the building and the length and types of egress components and the conditions in the egress route. Additionally, people will not necessarily use the shortest or most direct path to a safe place.

Impact of environment. The types of effects that the fire environment can have include psychological effects, such as turning back while exiting or deciding against choosing an egress path, and physiological effects, such as reduced visibility, incapacitation, or death.

Figure 6.1 illustrates the interrelationship between these factors (SFPE, 2003).

For some of these factors, generalized information is available that can be used to develop quantitative estimates. However, in others, a generalized

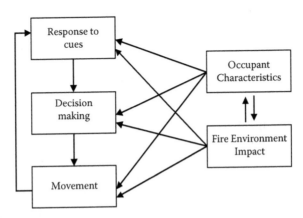

Figure 6.1 Interrelationship of human behavior factors. (From SFPE, 2003.)

Table 6.1 Addressing Human Behavior Factors

Subject	How Addressed
Occupant factors	Qualitative information available
Response to cues	Estimates must be developed from qualitative information
Decision	Qualitative information regarding the factors that influence decision making is available; case study information can be used to develop quantitative predictions
Impact of environment	Equations are available
Movement	Equations are available

Source: SFPE, 2003.

understanding does not exist, so estimates must be developed from qualitative information. Table 6.1 illustrates how each subject can be addressed.

OCCUPANT CHARACTERISTICS

The *SFPE Engineering Guide—Human Behavior in Fire* (2003) states:

> To predict human reactions and behaviors during a fire, the occupant characteristics of a building population need to be reviewed to identify the occupant group or groups that are important in the analysis. Using the list of occupant characteristics, a group or groups can be distinguished by their key characteristics. Not all characteristics are essential factors, but those that are critical and expected to influence the reaction and behavior of a group or groups should be noted.
>
> In performing the evacuation analysis, it may be possible to rely on a single defined occupant group that is recognized as the most critical and is conservatively characterized. However, it may also be necessary to perform additional analysis when several identified occupant groups in a given building are distinguished by their varying characteristics.

The *SFPE Engineering Guide—Human Behavior in Fire* (2003) provides the following occupant characteristics:

- **Population numbers and density.** The number of occupants and their density (number of occupants per unit area of floor space) in a building will influence the amount of time that it takes for people to egress through a building. The occupant load of a building will generally be identified by the requirements of a code, such as the *Life Safety Code* (2012), although published surveys of occupant loads can also be used. Surveys have generally found lower occupant loads than would be determined by using the factors in the *Life Safety Code*, so the factors identified in the *Life Safety Code* could be considered conservative, in that they provide an upper bound.

- **Familiarity with the building.** Familiarity with the building will influence which exits are used in the event of a fire. Occupants will tend to only use exits with which they are familiar. Therefore, if occupants frequently use a building (such as an office building), they would be more likely to use exit routes other than the one they used to come into the building. On the other hand, occasional users, such as would be expected in a shopping mall or theater, would be more likely to use the route that they used to enter the building. For example, occupants would be less likely to use an emergency exit in a shopping mall that consisted of a protected corridor if the occupants were not familiar with the route.

- **Distribution and activities.** The activities in which a person is engaged can influence the time that he or she spends before beginning to evacuate. If people are not evenly distributed throughout a building, then that will need to be considered when conducting a movement analysis.

- **Physical and cognitive ability.** Whether or not a person is disabled or impaired will influence his or her ability to sense cues, make decisions, and evacuate a building. Additionally, physical ability can influence a person's susceptibility to products of combustion.

- **Social affiliation.** If people are in a group, they will likely want to make sure that other people in the group are safe before they begin evacuation. For example, a family that has dropped their children off in a play area in a mall would likely not leave until they were sure that their children were safe.

- **Role and responsibility.** If people have a leadership role in a facility, then they will likely take charge in the event of an emergency.

- **Location.** The location of individuals in a building can influence the ability to sense fire cues, movement time, and selection of escape routes.

- **Commitment.** People are goal oriented, and will tend to want to finish activities that they have started. For example, people who have ventured to a shopping mall may want to complete their shopping before they leave.

- **Focal point.** In some facilities (e.g., theaters) people's attention may be focused at a particular point, which could influence their ability to sense cues in other parts of the building.

- **Occupant condition.** An occupant's condition (awake, asleep, intoxicated, etc.) can influence his or her ability to sense or respond to fire cues.

- **Culture.** Culture may influence human behavior; however, culture is unlikely to influence how human behavior is considered in a fire.

- **Age.** Age can affect other occupant factors.

RESPONSE TO CUES

Cues can come from four sources:

- Fire cues (e.g., seeing a fire or smoke, or smelling smoke)
- Building signaling systems (e.g., fire alarm or public address systems)
- Cues from other people (e.g., other occupants, building staff, or arrival of the fire department)
- Cues from building service interruptions (e.g., utility failures or breaking glass)

In general, people will not immediately recognize a fire cue as such. This is because fires are rare events, and most people do not have prior exposure to actual fires. Occupants will tend to go through the following process:

- Receive a cue (e.g., "I smell something unusual.")
- Recognize the cue (e.g., "It's smoke.")
- Interpret the cue (e.g., "That smoke might have come from a fire.")

This process will repeat as people are exposed to additional cues. Of the data available in the literature, a Canadian study (Proulx et al., 2001) is particularly noteworthy. This study identified that only 6% of study participants recognized the temporal-3 fire alarm signal as such. Therefore, it is unlikely that building occupants would begin evacuating when the only cue to which they were exposed was a fire alarm signal that used the temporal-3 pattern.

DECISION MAKING

There are a variety of actions that people may take after recognizing a fire cue: seeking additional information, searching for family (or other group) members, notifying others, fire fighting, beginning evacuation, or continuing with activities in which they were previously engaged.

Occupant characteristics will influence the types of decisions that people will make when exposed to fire cues. For example, people would likely take different actions upon first hearing a fire alarm if they were in an office building versus if they were in their home.

In general, initial fire cues will be regarded as ambiguous, since fires are rare events and are therefore unexpected by building occupants. When exposed to fire cues, people will frequently tend to attempt to gather additional information to confirm that there is a fire.

The time before movement begins is collectively known as pre-movement time. Pre-movement time includes the time for cues to be transmitted to occupants, for occupants to recognize cues, and the time for people to decide to evacuate. The pre-movement time can comprise as much or more time than it takes for people to egress once they have decided to do so.

Unfortunately, there are presently no generalized methods of estimating how long this time will be. Therefore, one must refer to information regarding past evacuations to derive such an estimate. Fahy and Proulx (2001) have compiled a database of the times necessary for pre-movement activities.

However, when applying information from previous evacuation studies, it is important to consider the context of the evacuation, for example, building type and whether it was an evacuation during an actual fire incident or a training exercise. Other contextual considerations include the weather conditions outside the building, the occupant characteristics, and building design factors.

MOVEMENT

From a purely physical standpoint, the time it takes to move to a location of safety is only a function of travel speed and distance. However, there are many factors that impact both travel speed and distance. Distance is a function of exit choice, which is affected by an occupant's familiarity with the structure, the availability of exits, and the degree of difficulty of a selected exit path. Travel speed is affected by a number of variables that include occupant mobility, crowding, presence of smoke, stair design, and widths of passages, corridors, and doors. Occupant training and the presence of trained staff to guide evacuation can also be factors.

After considering the above factors, the calculation of movement time can be accomplished using hand calculations or computer models. Hand calculations are based on a hydraulic analogy—that the flow rate of people through egress components is a function of the effective width of the component and the density of people using the component. Computer-based models either use a hydraulic approach or are based on individual behaviors of occupants.

The presence of smoke in an egress path can decrease the probability that occupants will move into an area or continue their evacuation. Also, smoke can reduce occupant walking speed. (See also the "Impact of the Environment" section.)

For an egress component, the capacity is a function of the component width. The entire width of a component is not available for use during egress. This is because of the tendency of people's bodies to sway as they move. Accordingly, the effective width of an egress component is the total width less a boundary layer on each side. See Table 6.2.

Table 6.2 Effective Widths

Element	Boundary Layer Width, in. (mm)
Stairways, walls, doors	6 (150)
Handrails	3.5 (90)
Corridors	8 (200)

Source: Gwynne, S., and Rosenbaum, E., Employing the Hydraulic Model in Assessing Emergency Movement, in *SFPE Handbook of Fire Protection Engineering*, Quincy, MA, 2008.

The total movement time is estimated as (Gwynne and Rosenbaum, 2008):

$$t = t_1 + t_2 + t_3$$

where t_1 is time for first person to reach controlling component, t_2 is time for population to move through controlling component, and t_3 is time for last person leaving controlling component to reach place of safety.

Movement speed is a function of the type of egress component (corridor, stairs, etc.) and the occupant density in the component. At densities less than 0.55 people per m² (0.05 people per ft²), the density is low enough that people move at an unimpeded speed. This speed can be calculated as (SFPE, 2003)

$$v = 0.85k$$

where v is speed, m/s (ft/min), and k is velocity factor (see Table 6.3).

As the occupant density increases above 0.55 people per m² (0.05 people per ft²), the movement speed begins to decrease. The movement speed at densities greater than 0.55 people per m² (0.05 people per ft²) can be calculated as follows:

$$v = k - akD$$

where a is constant, 0.266 m²/person (2.86 ft²/person), and D is density of occupant flow, people per m² (people per ft²).

Table 6.3 Velocity Factors

Egress Component		k (m/s)	k (ft/min)
Corridor, aisle, ramp, doorway		1.40	275
Stair Riser, mm (in.)	Stair Tread, mm (in.)		
190 (7.5)	254 (10)	1.00	196
272 (7.0)	279 (11)	1.08	212
165 (6.5)	305 (12)	1.16	229
165 (6.5)	330 (13)	1.23	242

Source: Gwynne, S., and Rosenbaum, E., Employing the Hydraulic Model in Assessing Emergency Movement, in *SFPE Handbook of Fire Protection Engineering*, Quincy, MA, 2008.

The density can be calculated by dividing the number of people in an egress component by the floor area in the component. Typical densities are less than 2 people per m² (0.19 people per ft²) (SFPE, 2003).

The speeds calculated from the equations above represent adult mobile people. Proulx (2008) provided data for people with mobility impairments as summarized in Tables 6.4 and 6.5.

Table 6.4 Mean Movement Speeds for People with Mobility Impairments on Horizontal Surfaces

Subject Group	Mean, m/s (ft/s)	Standard Deviation, m/s (ft/s)	Range, m/s (ft/s)
All disabled	1.00 (3.3)	0.42 (1.4)	0.10–1.77 (0.32–5.8)
With locomotion disability	0.80 (2.6)	0.37 (1.2)	0.10–1.68 (0.32–5.5)
No aid	0.95 (3.1)	0.32 (1.0)	0.24–1.68 (0.79–5.5)
Crutches	0.94 (3.1)	0.30 (1.0)	0.63–1.35 (2.1–4.4)
Walking stick	0.81 (2.7)	0.38 (1.2)	0.26–1.60 (0.85–5.3)
Walking frame or rollator	0.57 (1.9)	0.29 (1.0)	0.10–1.02 (0.32–3.4)
Without locomotion disability	1.25 (4.1)	0.32 (1.0)	0.82–1.77 (2.7–5.8)
Electric wheelchair	0.89 (2.9)	—	0.85–0.93 (2.8–3.1)
Manual wheelchair	0.69 (2.3)	0.35 (1.1)	0.13–1.35 (0.47–4.4)
Assisted manual wheelchair	1.30 (4.3)	0.34 (1.1)	0.84–1.98 (2.8–6.5)
Assisted ambulant	0.78 (2.6)	0.34 (1.1)	0.21–1.40 (0.69–4.6)

Source: Proulx, G., Evacuation Time, in SFPE Handbook of Fire Protection Engineering, Quincy, MA, 2008.

Table 6.5 Mean Movement Speeds for People with Mobility Impairments on Stairs

Subject Group	Mean, m/s (ft/s)	Standard Deviation, m/s (ft/s)	Range, m/s (ft/s)
Ascent			
With locomotion disability	0.38 (1.25)	0.14 (0.46)	0.13–0.62 (0.43–2.0)
No aid	0.43 (1.41)	0.13 (0.43)	0.14–0.62 (0.46–2.0)
Crutches	0.22 (0.72)	—	0.13–0.31 (0.43–0.98)
Walking stick	0.35 (1.15)	0.11 (0.36)	0.18–0.49 (0.50–1.6)
Rollator	0.14 (0.46)	—	—
Without disability	0.70 (2.30)	0.24 (0.79)	0.55–0.82 (1.8–2.7)
Descent			
With locomotion disability	0.33 (1.08)	0.16 (0.52)	0.11–0.70 (0.36– 2.3)
No aid	0.36 (1.18)	0.14 (0.46)	0.13–0.70 (0.43–2.3)
Crutches	0.22 (0.72)	—	—
Walking stick	0.32 (1.05)	0.12 (0.39)	0.11–0.49 (0.36–1.6)
Rollator	0.16 (0.52)	—	—
Without disability	0.70 (2.30)	0.26 (0.85)	0.45–1.10 (1.5–3.6)

Source: Proulx, G., Evacuation Time, in SFPE Handbook of Fire Protection Engineering, Quincy, MA, 2008.

The specific flow is a measure of the number of people who pass a fixed point in the egress point per unit time per unit distance of effective width (Gwynne and Rosenbaum, 2008). This is principally the same as a flux in fluid flow. The specific flow is calculated as follows:

$$F_s = Dv$$

where F_s is specific flow, persons/s-m (persons/min-ft); D is density of occupant flow, people per m² (people per ft²); and v is speed, m/s (ft/min).

By combining the equation for specific flow with the equations for movement speed, the specific flow can be calculated as follows:

$$F_s = (1 - aD)kD \text{ for } D > 0.55 \text{ people/m}^2 \ (0.051 \text{ people/ft}^2)$$

$$F_s = 0.85kD \text{ for } D < 0.55 \text{ people/m}^2 \ (0.051 \text{ people/ft}^2)$$

As a quadratic equation, the specific flow is subject to a maximum value; for greater densities the specific flow would decrease. This maximum specific flow occurs when the density is 1.9 people per m² (0.18 people per ft²) (SFPE, 2003). The maximum specific flows for a variety of egress components are shown in Table 6.6.

The total flow through an egress component is the specific flow multiplied by the effective width. Thus:

$$F_c = F_s W_e$$

where F_c is the total flow (people per unit time).

Substituting for F_s gives:

$$F_c = (1 - aD)kDW_e \text{ for } D > 0.55 \text{ people/m}^2 \ (0.051 \text{ people/ft}^2)$$

$$F_s = 0.85kDW_e \text{ for } D < 0.55 \text{ people/m}^2 \ (0.051 \text{ people/ft}^2)$$

Table 6.6 Maximum Specific Flows

Egress Component		Maximum Specific Flow, persons/s-m (persons/min-ft)
Corridor, aisle, ramp, doorway		1.32 (24.0)
Riser, mm (in.)	Tread mm (in.)	
190 (7.5)	254 (10)	0.94 (17.1)
272 (7.0)	279 (11)	1.01 (18.5)
165 (6.5)	305 (12)	1.09 (20.0)
165 (6.5)	330 (13)	1.16 (21.2)

Source: Gwynne, S., and Rosenbaum, E., Employing the Hydraulic Model in Assessing Emergency Movement, in *SFPE Handbook of Fire Protection Engineering*, Quincy, MA, 2008.

Generally, as people exit a building, they move through different egress components. For example, someone evacuating a high-rise building might leave his or her office and enter a corridor. The corridor might lead to exit stairs (which would be accessed through a door). As the building occupants traveled down the stairs, they would have to merge with people who enter the stairs from lower floors. Ultimately, the stairway would discharge through a door to the outside.

Each of these changes—movement from one egress component to another or merging with people already in an egress component or entering an egress component—will affect the time that it takes to evacuate the building. Flow rates through each of these transitions can be estimated by observing that the flow into a transition must equal the flow out of a transition (since people are neither created nor destroyed in the transition.) So, for merging flows:

$$F_{C_1} + F_{C_2} = F_{C_3}$$

where F_{C_1} and F_{C_2} are the total flows into the transition, and F_{C_3} is the total flow out of the transition. If the transition is simply from one egress component into another one (like going from a corridor through a door) then:

$$F_{C_1} = F_{C_2}$$

where F_{C_1} is the total flow into the transition, and F_{C_2} is the total flow out of the transition. Substituting for F_c and generalizing yields

$$\sum F_{s_{in}} W_{e_{in}} = F_{s_{out}} W_{e_{out}}$$

This equation can be used to determine the specific flow out of a transition given the sum of the products of the specific flows and effective widths of the egress components leading into the transition and the effective width of the egress component that leads from the transition. If, when using this equation, the calculated specific flow out of the transition exceeds the maximum specific flow (from Table 6.6), then the maximum specific flow should be used.

There are two different approaches that can be used to estimate the time necessary to move through a building. These methods only calculate movement time, and do not include the pre-movement time. The two methods are the first-order model and the second-order model (Gwynne and Rosenbaum, 2008).

To use the first-order model, the rate controlling element must first be determined, i.e., the component that has the lowest calculated flow. Then, four times must be determined and summed:

- The time for people to reach the rate controlling element
- The time for the population to move through the rate controlling element
- The time for the last person to leave the rate controlling element
- The time for the last person to reach a safe place

The second-order model involves calculating the flow of people through each element of the egress system. Both models are illustrated in Figure 6.2.

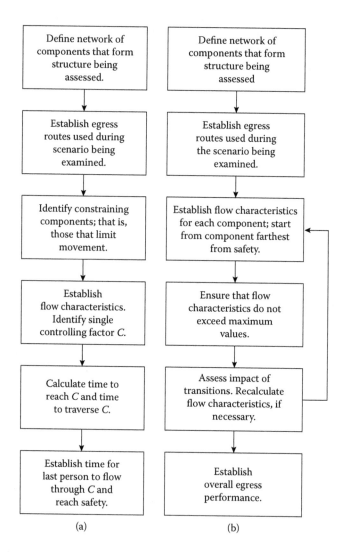

(a)

(b)

Figure 6.2 (a) First- and (b) second-order models. (From Rosenbaum and Gwynne, 2008.)

IMPACT OF THE ENVIRONMENT

When there is a possibility that building occupants will be exposed to fire products (smoke, heat, etc.), a common design objective is to avoid conditions that are harmful to occupants. This can be accomplished by several means (NFPA, 2012):

- Design such that fire effects will not reach any occupied space
- Design such that the smoke layer elevation would not descend below head height of any occupied space
- Design such that any occupied areas will be evacuated before the smoke layer descends below head height of any occupied space
- Conduct an analysis of the hazards in the fire-induced environment to determine whether they create an environment that could be incapacitating to people

The first three of these analyses are relatively straightforward, and will not be further addressed here. A fire-induced environment can create the following hazards to people (NFPA, 2007):

- Thermal injury
- Toxicity
- Reduction in visibility

Thermal Injury

Thermal injury can be caused by immersion of an individual in a heated atmosphere or by radiant transfer, e.g., from fires that are remote from an individual or from a hot smoke layer. The *SFPE Engineering Guide— Predicting 1st and 2nd Degree Skin Burns* (1999) provides methods of predicting whether a person would experience pain or first- or second-degree burns from exposure to thermal radiation. In general, as long as an individual is exposed to thermal radiation below a threshold of ~1.7 kW/m² (which is only about 50% more intense than the thermal radiation contained in sunlight), the person will not experience pain or burns, even for very long exposures. As the intensity of thermal radiation increases above this threshold, the time before an individual experiences pain or becomes injured decreases exponentially.

For exposure to hot gases at temperatures below those that could cause skin burns, the primary hazard is hyperthermia, where the threshold is about 80°C for humid air and 120°C for dry air (Purser, 2008). Since water is a combustion product of any fuel containing hydrogen (and all fuels of interest in building fire safety design will contain hydrogen), fire-induced environments should be considered as humid, and the lower threshold should be used. Purser (2008) provides a methodology for estimating whether hyperthermia could be expected that is a function of temperature and exposure time.

Burns to the respiratory tract from inhaling hot gases are also possible; however, it is likely that an individual would have already been injured by skin burns before respiratory tract burns occur. Therefore, consideration of whether respiratory tract burns could occur is typically not necessary (Purser, 2008).

Toxicity

If people are exposed to toxic gases at a concentration sufficient to cause harm, they could become incapacitated or unable to continue to affect their own evacuation without harm. If they continue to be exposed to hazardous concentrations of fire products, death could eventually result. When developing building designs that are intended to achieve life safety goals and people could be exposed to fire products, avoidance of incapacitation is generally the design objective.

Experiments involving exposure of fire products to animals provide the basis for fire product toxicity data. For exposures to concentrations that are lower or higher than used in laboratory experiments, the toxicity can be adjusted using Haber's rule (Purser, 2008), which states that

$$W = C \times T$$

where W is a constant dose (such as an incapacitating dose), C is the concentration of the fire product, and T is the time of concentration.

There are two approaches that can be used to estimate the effect of exposure to products of combustion. The first approach is to estimate the mass loss of fuel. Most of the data for this type of analysis come from well-ventilated burning conditions (Purser, 2008), so their use in cases of under-ventilated burning may not be appropriate.

Purser (2008) notes that LC_{50} (the exposure that was lethal for 50% of the test animals) has been found to range from 150 to 1,800 g-m^{-3}-min (the units reflect mass loss per cubic meter of room volume multiplied by time of exposure). It is noteworthy that this value is the exposure that was lethal to half of the animals exposed. Therefore, even selecting the lowest value in this range would not be conservative for design, since as many as 50% of the people exposed to this condition could perish, and incapacitation would be expected at lower concentrations than the lethal concentration.

To account for incapacitation and for people who are more sensitive to fire products than the average population (such as asthmatics and the very old), a NIST report (Gann et al., 2001) suggests a design value of 6 g-m^{-3} for well-ventilated burning and 3 g-m^{-3} for post-flashover burning conditions for a 5 min exposure, which equates to 30 and 15 g-m^{-3}-min, respectively. The report also estimates the uncertainty in these values as being a factor of 2, which suggests dividing these values by 2 if used for tenability analysis in design.

The second approach is to estimate the fractional effective dose (FED), which involves estimating the yields of individual species (CO, CO$_2$, HCN, etc.) and calculating their effect on people exposed to them. The fractional effective dose is based on the belief that the effects of toxic products on people are linearly cumulative, e.g., that if one is 75% incapacitated from CO and 25% incapacitated by HCl, then that person is 100% incapacitated.

In general, CO and CO$_2$ yields in fires can be calculated with the most confidence, and in most cases they are also the toxicants of primary concern. The yields of CO and CO$_2$ are strongly influenced by burning conditions, with CO yields low in cases where there is sufficient oxygen to completely burn the fuel, and higher in cases of smoldering combustion and ventilation-controlled (e.g., post-flashover) burning.

When using the FED, for each species the $C \times T$ dose exposed is divided by the $C \times T$ dose that gives the effect of interest, e.g., incapacitation. These fractional effective does are then summed for each of the species to which people are exposed.

Purser (2008) presents the following equation for calculation of FED:

$$F_{IN} = \left[(F_{ICO} + F_{ICN} + FLC_{irr}) \times VCO_2 + FED_{Io} \right] \text{ or } F_{ICO_2}$$

where F_{IN} is fraction of an incapacitating dose of all asphyxiating gases, F_{ICO} is fraction of an incapacitating dose of CO, F_{ICN} is fraction of an incapacitating dose of HCN, FLC_{irr} is fraction of irritant dose, VCO_2 is multiplication factor for CO$_2$-induced hyperventilation, FED_{Io} is fraction of an incapacitating dose of low-oxygen hypoxia, and F_{ICO_2} is fraction of an incapacitating dose of CO$_2$.

Purser (2008) gives the following equations for calculation of the individual fractional effective doses:

$$F_{ICO} = \frac{8.2925 \times 10^{-4} \times [CO]^{1.036}}{30},$$

where [CO] is concentration of CO, expressed in parts per million;

$$F_{ICN} = \frac{\exp\left(CN/43 \right)}{220},$$

where CN is concentration of HCN in parts per million added to the concentration of other nitriles minus the concentration of NO$_2$;

FLC_{irr}

which is the fraction of the incapacitating dose from all incapacitating products (HCl, HBr, etc.);

$$VCO_2 = \exp\left(\frac{[CO_2]}{5}\right),$$

where $[CO_2]$ is concentration of carbon dioxide in percent;

$$FED_{I_O} = \left\{\exp\left[8.13 - 0.54\left(20.9 - [O_2]\right)\right]\right\}^{-1},$$

where $[O_2]$ is concentration of oxygen in percent; and

$$F_{I_{CO_2}} = \left\{\exp\left[6.1623 - 0.5189 \times [CO_2]\right]\right\}^{-1},$$

where $[CO_2]$ is the concentration of CO_2 in percent.

The equations for FED and the components of FED are based on a 1 min exposure. For exposures to constant concentrations of fire products, the FED can be determined by multiplying the value determined using the above equations by the exposure time in minutes. For exposures where the concentrations vary with time, the total FED can be calculated by discretizing the exposure (determining the average exposure at each 1 min interval and summing the FEDs determined for each 1 min interval).

The mechanism of harm for incapacitating products is different from that for asphyxiating products. Most fire safety analyses will focus on asphyxiating gases (and generally just CO and CO_2). The impact from asphyxiating gases is discussed in Purser (2008).

As with calculations involving tenability on the basis of mass loss, calculations of FED should consider more sensitive populations. Purser (2008) suggests that fractional effective doses greater than 0.1 should be avoided. Gann et al. (2001) take a different approach to accounting for sensitive populations, where they provide incapacitating doses for a fixed time exposure. Where the concentration of a toxicant varies from the values provided by Gann et al. (2001), the authors recommend scaling based on C^2T, as opposed to Haber's rule, which suggests $C \times T$.

The FED is most appropriate for asphyxiating or narcotic gases (e.g., CO, CO_2, HCN, and reduced oxygen). For irritating gases, such as HCl, which have a different mechanism of harm, a fractional irritant concentration (FIC) is more appropriate. The effect of exposure to irritating gases is acute, rather than cumulative. If a person is exposed to an atmosphere for which the FIC is greater than 1, the person would be expected to become immediately incapacitated (Purser, 2008).

Visibility

The third way that a fire environment can influence human behavior is by causing a reduction in visibility. While visibility does not in itself cause physiological harm, reduction in visibility can hinder egress and limit actions of occupants. In most cases, visibility thresholds are the failure points that are reached first. Reduced visibility can impact human behavior in many ways:

- Reduce the likelihood that people will move through an egress route
- Reduce movement speeds in egress routes
- Reduce wayfinding ability

Bryan (2008) found that as the visibility in smoke decreased, the likelihood that people would choose to move through an egress route deceased. Additionally, Bryan found that as visibility distances in egress routes decreased, the likelihood that people would turn around increased. Table 6.7 shows the visibility distances for British and U.S. populations for which people moved through smoke. Table 6.8 shows the visibility distances for British and U.S. populations at which people turned back in egress routes.

Jin (2008) studied the impact of smoke on movement speed. He found that movement speeds decreased in smoke, particularly when the smoke was of an irritating nature. Figure 6.3 shows Jin's findings. Jin burned wood cribs to create irritating smoke and kerosene to create nonirritating smoke. When doing a design analysis, if it is uncertain whether or not

Table 6.7 Visibility Distances for British and U.S. Populations When People Moved through Smoke

Visibility Distance, m (ft)	UK Sample Population, %	U.S. Sample Population, %
0–0.6 (0–2)	12.0	10.2
0.9–1.8 (3–6)	25.0	17.2
2.1–3.7 (7–12)	27.0	20.2
4.0–9.1 (13–30)	11.0	31.7
9.4–11 (31–36)	3.0	2.2
11–14 (37–45)	3.0	3.7
14–18 (46–60)	3.0	7.4
>18 (60)	17.0	7.4

Source: Bryan, J., Behavioral Response to Fire and Smoke, in SFPE Handbook of Fire Protection Engineering, National Fire Protection Association, Quincy, MA, 2008.

Table 6.8 Visibility Distances for British and U.S. Populations at which People Initiated Turn-Back Behavior

Visibility Distance, m (ft)	UK Sample Population, %	U.S. Sample Population, %
0–0.6 (0–2)	29.0	31.8
0.9–1.8 (3–6)	37.0	22.3
2.1–3.7 (7–12)	25.0	22.3
4.0–9.1 (13–30)	6.0	17.6
9.4–11 (31–36)	0.5	1.3
11–14 (37–45)	1.0	0
14–18 (46–60)	0.5	4.7
>18 (60)	1.0	0

Source: Bryan, J., Behavioral Response to Fire and Smoke, in *SFPE Handbook of Fire Protection Engineering*, National Fire Protection Association, Quincy, MA, 2008.

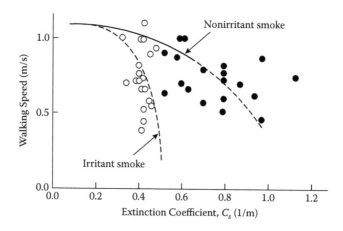

Figure 6.3 Walking speed in fire smoke. (From Jin, T., Visibility and Human Behavior in Fire Smoke, in *SFPE Handbook of Fire Protection Engineering*, National Fire Protection Association, Quincy, MA, 2008.)

smoke would be irritating, then it should be assumed that the smoke would be irritating.

Lastly, reduced visibility can decrease wayfinding, decreasing the likelihood that people will be able to find their way to an exit. When considering the impact of smoke on an occupant's ability to wayfind, the ability to view lighted exit signs or other objects is generally used as a design objective. Smoke decreases visibility through two mechanisms: (1) smoke particles block light from a sign before it reaches an occupant, and (2) light from sources other than a sign is scattered, which can overwhelm the light from an exit sign (Jin, 2008). It is only possible to distinguish an object from its surroundings when the contrast between the sign and its surroundings is greater than some threshold.

In general, a threshold of 2% brighter is considered the minimum contrast threshold necessary to discern a sign from its surroundings (Jin, 2008; Mulholland, 2008). Therefore, the visibility distance of an object is the distance at which the contrast between an object and its surroundings is reduced to 2%. However, most studies of the visibility of exit signs in smoke have used test subjects rather than precise measurements of contrast using a photometer (Mulholland, 2008).

Mulholland (2008) provides the following equations for visibility distance in smoke:

$$K \times S = 8 \text{ (light emitting signs)}$$

$$K \times S = 3 \text{ (light reflecting signs)}$$

where K is extinction coefficient in smoke (m^{-1}), and S is visibility distance (m).

In the experiments used to develop the equations above, subjects were not exposed to the fire smoke, which means that they are only applicable for nonirritating smoke.

Jin (2008) also provides a method of estimating the visibility distance of exit signs that is comparable to Mulholland's. While Jin's methods are based on visibility of exit signs, Jin suggests that the methods for calculating visibility distances of exit signs may be applicable to other objects, such as walls, floors, doors, and stairways.

As with Mulholland's method, Jin's equations are based on nonirritating smoke. Mulholland's method is generally used in practice. For irritating smoke, Jin found that visibility distance decreased precipitously once the smoke density reached a certain level.

Some references provide performance criteria in the form of limiting extinction coefficients or optical densities. Babrauskas (1979) suggested a visibility distance of 1.67 m, which was based on ensuring sufficient visibility for an occupant to exit the room of fire origin. Babrauskas's suggested visibility was used for evaluating performance in mattress fires, and he assumed that once people exited the room of fire origin, there would be sufficient visibility in the corridor.

Jin (1981) suggested a minimum visibility distance of 3 to 5 m for people who are familiar with a building and 15 to 20 m for people who are not familiar with a building. These values were developed by determining the smoke density that reduced walking speeds to those that were found when subjects were asked to move through a darkened (but not smoke-filled) corridor.

Rasbash (1975) suggested a minimum visibility distance of 10 m. A questionnaire was administrated to fire service personnel asking them to estimate the visibility distances at which people would (1) move into a smoke-filled corridor and (2) turn around if already in a smoke-filled corridor. Analysis of the data showed no correlation between the smoke visibility and willingness to enter a smoke-filled corridor. However, the data showed that at visibility distances less than 10 m, the percentage of people in a smoke-filled corridor who would turn around increased above 10%. When using these data, it should be noted that no actual measurements were taken (the data were based on the recollections of fire service personnel). To wit, even the author noted that the 10 m threshold was arbitrary.

MODELS

Models are frequently used to perform calculations involving the concepts identified in this chapter. The models used are generally computer-based applications of the concepts identified in this module. Prior to using a human behavior model, the modeler should identify how human behavior concepts are modeled so that he or she can determine the level of confidence to place on model results. Kuligowski and Gwynne (2005) provide suggestions for what a modeler should look for when selecting a human behavior model. These include:

- Identify project objectives
- Understand model origin, i.e., the available expertise at the time the model was developed and whether the model has been validated
- Model characteristics

REFERENCES

Babrauskas, V., Technical Note 1103, National Bureau of Standards, Gaithersburg, MD, 1979.

Bryan, J., *Implications for Codes and Behavior Models from the Analysis of Behavior Response Patterns in Fire Situations as Selected from the Project People and Project People II Study Programs*, National Bureau of Standards, Gaithersburg, MD, 1983.

Bryan, J., Human Behavior in Fire—The Development and Maturity of a Scholarly Study Area, in *Human Behavior in Fire: Proceedings of the First International Symposium*, University of Ulster, Derry, UK, 1998.

Bryan, J., Behavioral Response to Fire and Smoke, in *SFPE Handbook of Fire Protection Engineering*, National Fire Protection Association, Quincy, MA, 2008.

Canter, D., *Studies of Human Behaviour in Fire: Empirical Results and Their Implications for Education and Design*, Building Research Establishment, UK, 1985.

Fahy, R., and Proulx, G., Toward Creating a Database on Delay Times to Start Evacuation and Walking Speeds For use in Evacuation Modeling, presented at Conference Proceedings—2nd International Symposium on Human Behavior in Fire, Interscience, London, 2001.

Gann, R., et al., *International Study of the Sublethal Effects of Fire Smoke on Survivability and Health (SEFS): Phase I Final Report*, NIST Technical Note 1439, National Institute of Standards and Technology, Gaithersburg, MD, 2001.

Gwynne, S., and Rosenbaum, E., Employing the Hydraulic Model in Assessing Emergency Movement, in *SFPE Handbook of Fire Protection Engineering*, Quincy, MA, 2008.

Jin, T., Studies of Emotional Instability in Smoke from Fire, *Journal of Fire and Flammability*, 12, 130, 1981.

Jin, T., Visibility and Human Behavior in Fire Smoke, in *SFPE Handbook of Fire Protection Engineering*, National Fire Protection Association, Quincy, MA, 2008.

Kuligowski, E., and Gwynne, S., What a User Should Know When Selecting an Evacuation Model, *Fire Protection Engineering*, Fall 2005.

Mulholland, G., Smoke Production and Properties, in *SFPE Handbook of Fire Protection Engineering*, National Fire Protection Association, Quincy, MA, 2008.

NFPA, *Life Safety Code*, NFPA 101, National Fire Protection Association, Quincy, MA, 2012.

Proulx, G., Evacuation Time, in *SFPE Handbook of Fire Protection Engineering*, Quincy, MA, 2008.

Proulx, G., et al., *Fire Alarm Signal Recognition*, Internal Report 828, IRC-IR-828, National Research Council of Canada, Ottawa, 2001.

Purser, D., Assessment of Hazards to Occupants from Smoke, Toxic Gasses and Heat, in *SFPE Handbook of Fire Protection Engineering*, National Fire Protection Association, Quincy, MA, 2008.

Rasbash, D., Smoke and Toxic Gas, *Fire*, 59, 735, 175–179, 1966.

SFPE, *SFPE Engineering Guide—Predicting 1st and 2nd Degree Skin Burns*, Society of Fire Protection Engineers, Bethesda, MD, 1999.

SFPE, *SFPE Engineering Guide—Human Behavior in Fire*, Society of Fire Protection Engineers, Bethesda, MD, 2003.

SFPE, *SFPE Engineering Guide to Performance-Based Fire Protection*, National Fire Protection Association, Quincy, MA, 2007.

Wood, G., A Study of Human Behaviour in Fires, in *Fires and Human Behaviour*, John Wiley & Sons, Chichester, UK, 1980.

Chapter 7

Detection and Suppression System Design

INTRODUCTION

Fire suppression and alarm systems play a key role in protecting people and property from fire. Being able to analyze the impact that these systems have on protecting people and property plays an important part in determining the level of safety they provide. Estimating detection response is the first step in analyzing the impact of these systems. Fire detection must occur before it is possible to activate suppression. For this reason, fire detection becomes an important element in performance-based design.

Detection systems take many forms and initiate most of the activities that affect the basic elements of fire response. For example, the following are all potential means of detection:

- A building occupant sees or smells a fire, and then notifies other occupants and calls the fire department (manual detection, notification, and suppression).
- A smoke detector activates an alarm system, notifying building occupants and the fire department (automatic detection and notification, manual suppression).
- A sprinkler activates, triggering a water flow switch, which in turn activates a fire alarm system, notifying building occupants and the fire department (automatic detection, notification, and suppression).

In each example, the type of detection system used affects the level of fire safety provided.

Calculation methodologies exist for estimating activation of smoke and heat detection. Each has its limitations. In some cases, detection response must be estimated by making assumptions. For example, equations don't exist for estimating detection by a person in the room of fire origin.

To evaluate the performance of a suppression system, it is necessary to determine when it is activated, and following activation, the effect that it will have on the fire.

This chapter presents information on methods used to estimate activation times of detection systems. Information is also provided on determining the impact of suppression systems on fire size.

DETECTION SYSTEM DESIGN

Fire detection occurs when there becomes an awareness of a fire. This can occur manually (i.e., someone seeing or smelling a fire) or automatically (i.e., by some type of detection system). Once the fire is detected, actions can be performed to mitigate the hazard or minimize resultant losses.

TYPES OF DETECTORS

The methodologies used to calculate detection depend on the type of detector. There are an assortment of detectors (NFPA, 2013), which are classified by the fire signatures they detect and their mechanism of operation. The archetypal models are:

- Manual detection
- Heat detectors
  - Fixed temperature
  - Rate of rise
  - Rate compensating
- Smoke detectors
  - Beam (obscuration)
  - Ionization
  - Photoelectric (light scattering)
  - Air sampling
- Radiant energy detectors
  - Ultraviolet
  - Visible
  - Infrared
  - Spark/ember detectors

Other detection technologies available include video imaging detection, gas detection and thermal wire detection. These are not addressed in this chapter.

This chapter will present calculation methodologies and application usage to determine detector activation. Some of these methodologies can also be used to predict activation of sprinkler systems, which are activated in the same manner as fixed temperature heat detectors. This chapter presents methodologies for heat and smoke detectors, but does not address those for radiant energy.

Manual Detection

People have the ability to detect fires with sight, sound, and smell. These senses can be quite acute. For example, a human being is capable of detecting the smell of certain chemicals to the sub-parts per billion range (Ruth, 1986). This detection level can vary considerably from person to person. Studies have shown (Walker et al., 2003) that the variability of scent detection can approach three orders of magnitude. The senses of sight and sound can also have variations across the population.

In addition to the wide variation in senses, attention plays an important role in the ability to detect a fire. While automatic detectors constantly assess for fire conditions, humans can be inattentive or become distracted. Distractions may include physical or mental activities, or temporary impairments such as injuries or a drug or alcohol impairment.

Even if a person is able to detect the signature of a fire and is not distracted, he or she may still not react to a fire. The perception of threat also plays a role. If the person perceives the fire to come from an innocuous source, such as burning toast, he or she may disregard the threat and react as if he or she did not detect it. Other people also play a role in the perception of threat. One study (Bryan, 2008) showed threat perception can change the ability to detect and react to a fire. When only one person was in the room, students reported seeing smoke 75% of the time. When two people who don't react joined the student, only 10% of the students reported seeing smoke. The additional people were able to change the perception of threat that the students experienced. Response to fire cues is discussed more fully in Chapter 6.

The combination of the variability of detecting a fire and the challenges associated with attentiveness and perception of threat makes predicting manual detection with any reliability next to impossible. Common sense would indicate that people in the immediate area of origin with direct line of sight to the fire would detect and react to a fire. This would be reasonable for most situations, but may not be appropriate under all circumstances.

Engineering judgment is required to estimate manual detection. Table 7.1 shows various means occupants used to detect a fire.

Heat Detectors

Heat detectors detect fires based on significantly elevated temperature. They can further be broken down into fixed temperature, rate of rise, and rate compensating detectors (NFPA, 2013). Fixed temperature detectors activate when the sensor temperature reaches a fixed temperature threshold. For purposes of determining activation times, automatic sprinklers can be treated as fixed temperature heat detectors because they use a fixed temperature sensing element. Rate of rise detectors activate when they sense a

Table 7.1 Means of Awareness of the Fire

Means of Awareness	Participants	Percent
Smelled smoke	148	26.0
Notified by others	121	21.3
Noise	106	18.6
Notified by family	76	13.4
Saw smoke	52	9.1
Saw fire	46	8.1
Explosion	6	1.1
Felt heat	4	0.7
Saw/heard fire department	4	0.7
Electricity went off	4	0.7
Pet	2	0.3
Total	569	100.0

Source: Bryan, J., Behavioral Response to Fire and Smoke, in *SFPE Handbook of Fire Protection Engineering,* 4th ed., National Fire Protection Association, Quincy, MA, 2008.

certain rate of temperature rise, such as 8°C/min (15°F/min). Rate compensating detectors activate based on the surrounding air temperature reaching a fixed temperature. They are designed to have a low thermal lag resulting in a sensitive fixed temperature detector.

Fixed Temperature Detectors

Fixed temperature detectors activate when the sensing element reaches a certain temperature. The temperature of the sensing element can be determined by calculating the heat transfer to this sensing element.

The heat transfer to the sensing element comes from four sources (Schifiliti et al., 2008; McGrattan et al., 2013):

- Conduction through the attachment point
- Convection from the surrounding gas
- Cooling by water droplets in the gas stream by previously activated sprinklers
- Radiation from the fire and surrounding gas

Of these four sources, convection from the surrounding gas is generally the dominant mode of heat transfer, and the other sources are often neglected. Convection is calculated based on the temperature and velocity of the gas around the sensing element that results from the ceiling jet and fire plume. Radiation heat transfer is comparatively small due to the typical size and location of sensing elements, as well as the gas temperatures at activation. Cooling by water droplets may play a significant role if there are sprinklers that have previously been actuated; however, it can be ignored when computing the actuation time for the first sprinkler or detector.

Assuming the total heat transfer to the sensing element is dominated by convection, the heat transfer to the sensing element can be expressed as

$$\dot{Q}_{convection} = hA(T_g - T_d) \tag{7.1}$$

where $\dot{Q}_{convection}$ is rate of convective heat transfer (kW), h is convective heat transfer coefficient for the detector (kW/m² · °C), A is surface area of the detector element (m²), T_g is temperature of fire gases at the detector (°C), and T_d is temperature rating or set point of the detector (°C).

Based on the size and composition of a typical sensing element, it can be treated as a lumped mass, m. The temperature change, dT_d, of the element can be expressed as

$$\frac{dT_d}{dt} = \frac{\dot{q}}{mc} \tag{7.2}$$

where t is time (s), $\dot{q}$ is rate of heat flow into mass (kW), m is mass of detector element (kg), and c is specific heat of detector element (kJ/kg · °C).

Estimating the mass, specific heat, and surface area of the sensing element may be challenging due to the size and shape. Calculating the convective heat transfer coefficient can be challenging due to the dependence on the gas velocity and element shape. Heskestad and Smith (1976) developed a simple test known as a plunge test that can be used to quantify the combination of these terms for automatic sprinklers and heat detectors. The test is used to determine a time constant for detector response. The time constant is defined as follows:

$$\tau = \frac{mc}{hA}. \tag{7.3}$$

This time constant is valid only for the tested gas velocity. To compensate for this, the concept of the response time index (RTI) was introduced. It is defined as

$$\tau u^{1/2} \approx \tau_0 u_0^{1/2} = RTI \tag{7.4}$$

where τ is time constant (s), u is velocity of fire gases (m/s), τ_0 is time constant at tested velocity (s), and u_0 is tested velocity.

RTI is used to describe the sensitivity of a heat detector or sprinkler. The RTI for a fast response sprinkler is 50 s$^{1/2}$ m$^{1/2}$ or less; a standard response sprinkler is 80 s$^{1/2}$ m$^{1/2}$ or more (NFPA 13, 2013). Manufacturers do not

Table 7.2 Time Constants

Listed Spacing (m)	Underwriters Laboratories Listed Temperature Rating						Factory Mutual Listed Detectors (all temperatures)
	53.3°C	57.2°C	62.8°C	71.1°C	76.7°C	91.1°C	
3.05	400	330	262	195	160	97	196
4.57	250	190	156	110	89	45	110
6.1	165	135	105	70	52	17	70
7.62	124	100	78	48	32	—	48
9.14	95	80	61	36	22	—	36
12.19	71	57	41	18	—	—	—
15.24	59	44	30	—	—	—	—
21.34	36	24	9	—	—	—	—

Source: NFPA, National Fire Alarm Code, NFPA 72, National Fire Protection Association, Quincy, MA, 2013.

typically publish values for RTI or time constants in product literature. For heat detectors, the listed spacing for a detector can be used to determine an equivalent RTI (see Table 7.2).

Combining all these equations into one yields the following for the heat transfer to the sensing element:

$$\frac{dT_d}{dt} = \frac{u^{1/2}(T_g - T_d)}{RTI} \tag{7.5}$$

This model neglects radiation and conduction, and it also does not account for latent heat required to melt the sensing link (for thermal link devices), nor does it account for variations in the heat transfer coefficient due to changes in orientation and velocity. Several researchers have proposed additional terms to account for these simplifications. A frequently encountered variation is to add the conduction back into the model (Heskestad and Bill, 1988). The modified model is as follows:

$$\frac{dT_d}{dt} = \frac{u^{1/2}(T_g - T_d)}{RTI} + \frac{C(T_g - T_d)}{RTI} \tag{7.6}$$

The C parameter represents conductive losses. Evans and Madrzykowski (1981) determined that conductive losses, for a particular sprinkler tested, accounted for about 20% or less of the time constant. They evaluated a two-parameter model against the simplified model above, but found that the differences between the two were within the accuracy and repeatability of the sprinklers and methods used.

The computer model Fire Dynamics Simulator (FDS) includes the conduction variation as well an empirical means of assessing the effect of water cooling (McGrattan et al., 2013).

The fire related unknowns in these models are the gas temperature and gas velocity. Historically, these quantities have been determined through ceiling jet and plume correlations (Alpert, 2008; Heskestad, 2008).

One set of correlations that have been used for detection were developed by Alpert (1972). The correlations are used to calculate ceiling jet gas temperatures and velocities as follows:

$$T_g - T_a = \frac{5.38 \dot{Q}^{2/3}}{H^{5/3} \left(r/H \right)^{2/3}} \text{ where } r/H > 0.18 \text{ or}$$

(7.7)

$$T_g - T_a = \frac{16.9 \dot{Q}^{2/3}}{H^{5/3}} \text{ where } r/H \le 0.18$$

$$u = \frac{0.197 \left(\dot{Q}/H \right)^{1/3}}{\left(r/H \right)^{5/6}} \text{ where } r/H > 0.15 \text{ or}$$

(7.8)

$$u = 0.947 \left(\frac{\dot{Q}}{H} \right)^{1/3} \text{ where } r/H \le 0.15$$

where T_a is ambient temperature (°C), $\dot{Q}$ is total heat release rate (kW), r is radial distance from plume axis to detector element (m), and H is ceiling height (m).

These correlations were developed for steady-state fires on flat, unconfined ceilings (no hot gas layer effects). These correlations can be used for growing fires as long as the fire grows slow enough to appear to be steady or quasi-steady (Zalosh, 2003). This approximation holds when

$$\frac{H}{u\dot{Q}} \frac{d\dot{Q}}{dt} \ll 1$$

(7.9)

The maximum temperatures and velocities of the ceiling jet occur near the ceiling. These temperatures and velocities fall off exponentially with the distance below the maximum. The thickness of the ceiling jet can be approximated as 10 to 12% of the fire-to-ceiling height (Alpert, 2008). The closer the sensing element is to the maximum temperature and velocities, the more reasonable the prediction (SFPE, 2002).

These correlations and the simplified detector model can be combined and integrated to determine the sensing element temperature as a function of time. This approach forms the basis of the computer program DETACT-QS (Evans and Stroup, 1985).

The preceding correlations are generally valid for larger spaces or for smaller spaces with short activation times. When the space is small or there is a delay in detection, the effects of a hot gas layer influence the temperatures and velocities of the ceiling jet. Cooper has developed correlations for confined ceiling jets in a hot gas layer. These correlations form the basis of the detection routines in CFAST (Peacock et al., 2005) and LAVENT (Davis and Cooper, 1989). LAVENT has the added benefit of being able to calculate activation times for detectors located outside the maximum temperature region of the ceiling jet. More recently, these correlations have been incorporated into the model jet (Davis, 1999).

In addition to correlations, the temperature and velocity of the gas can be calculated through the use of computational fluid dynamics (CFD) programs. The advantage of using a CFD program is that it allows additional physics to be included that cannot be replicated by correlations. This includes the effects of supply and exhaust points for ventilation systems and the effects of sprinkler discharge, which can disrupt the ceiling jet. The biggest disadvantage to using a CFD program is the additional computation time and skill level to obtain the input data, implement and run the models, and interpret results. Typically, running a CFD program is several orders of magnitude more computationally intensive than the correlation-based approach (hours or days versus seconds).

Rate of Rise

Rate of rise detectors activate when they sense a certain rate of temperature rise, such as 8°C/min (15°F/min). The mechanisms used to identify this rate of rise are different than those of a fixed temperature detector (Schifiliti et al., 2008). Typically the lumped mass methodology is used with the ceiling jet and plume correlations, except the equations are solved for the temperature rate of change of the sensing element.

Rate Compensated

Rate-compensated detectors are sensitive fixed temperature detectors. They are designed to compensate for the thermal lag in the detector and activate when the gas temperature reaches a specific temperature. The mechanisms used to determine this are more complicated than for a fixed temperature detector (Schifiliti et al., 2008).

Nam et al. (2004) studied fixed temperature and rate-compensated detectors to determine if the standard plunge-type test would give satisfactory results for real-world situations. The results of this study showed that the plunge test results were adequate for the fixed temperature and lower temperature rate-compensated detectors. High-temperature rate-compensated detectors activated earlier than the lumped mass assumption predicted, and so the assumption still provides a conservative estimate.

Current analysis techniques for dealing with the compensated portion of the rate-compensated detector are limited. Therefore, it is recommended to treat the detector as a fixed temperature detector. Based on how it operates, the lumped mass methodology based on a fixed temperature will yield conservative results (predict activation later than what would actually happen) for a rate-compensated detector.

Smoke Detectors

Smoke detectors are devices that respond to the presence of particulates (smoke) in air. They generally offer earlier response times to fires than heat detectors. However, prediction of smoke detector actuation is generally more difficult than prediction of heat detector activation. Particle color, number density, mass concentration, and particle size average and distribution are additional factors to consider in smoke detection (Schifiliti et al., 2008) that are not required for heat detectors.

Prediction of smoke detector operation requires estimating smoke concentrations exposing a detector, smoke entry into the detector, and activation of the alarm mechanism. Several potential methods to predict smoke detection activation are available, and each has its benefits and limitations.

Ionization and photoelectric smoke detectors are addressed in this chapter. Ionization detectors operate by ionizing the air between two electrodes in a chamber, allowing a small current to flow between electrodes. Smoke enters the chamber and interrupts the flow of current, causing detection. Photoelectric (light scattering) detectors contain a light source and light receiver (photocell). The receiver does not normally receive any light from the light source. When particles (smoke) are introduced, the light scatters in random directions. The scattered light is detected by a photocell, which is normally outside the line of sight of the light source. Other types of smoke detectors include beam, air sampling detectors, and video imaging detectors.

Detection Criteria

Smoke can vary in particle concentration, average size, color, and size distribution. Different combustible materials will produce smoke with different variations in properties when burned. For a given material, the properties and quantities of smoke produced depend on the combustion conditions (flaming, smoldering, pyrolysis) (Mulholland, 2008).

Table 7.3 Smoke Concentration at Detector Activation

Material	Smoke Concentration (mg/m³)	
	Ionization	*Photoelectric*
Wood	4.2	13.0
Cotton	0.5	6.9
Polyurethane	43.4	43.4
PVC	86.8	86.8

The smoke properties affect the ability of a detector to detect a fire. Each type of detector evaluates the smoke particles differently, so the type of detector also has an effect on detection. Table 7.3 shows the effect that various flaming materials have on the smoke concentration needed to activate an ionization or photoelectric detector. This table was adapted from NFPA 72 (2013) using a specific extinction coefficient of 8.7 m²/g, as suggested by Mulholland and Croarkin (2000).

One of the important quantities for smoke detection is the size of the particle. Figure 7.1 shows detector response functions of an ionization detector and a photoelectric detector for various particle sizes. The response functions estimate the electronic signal response of the two detection technologies to smoke particles at various sizes.

These response functions are specific to individual detectors (Mulholland, 2008). They will change based on the detector design features, such as inlets and smoke chamber properties and materials used. One manufacturer (Qualey and Penny, 2005) found significant differences in the response functions between two similar photoelectric detectors when the smoke chamber was constructed of different types of plastic.

The photoelectric response function (Mulholland, 2008) will also change depending on the light source, scattering angle, and scattering volume. With the right combination of light source and scattering angle, light scattering detectors (e.g., air sampling detector) can be much more sensitive to smaller particle sizes.

The smoke particle size is governed by the combustion conditions for a given material (Mulholland, 2008). Small particles are produced by flaming fires, and larger particles are produced by smoldering fires or pyrolyzing materials. Based on the detector response functions detailed in Figure 7.1, the ionization detector would detect a flaming fire with small particles sooner than the photoelectric detector specified. Conversely, the photoelectric detector in Figure 7.1 would be better at detecting a smoldering fire with large particles than the ionization detector.

One complicating factor is in providing a reference measurement of the smoke concentration at detector activation. The smoke concentration measured in tests depends on the device used to measure it. Smoke concentrations can be estimated using light attenuation or a reference ionization chamber

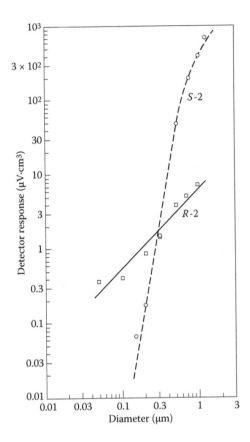

Figure 7.1 Detector response function plotted versus particle size for detectors S-2 (light scattering) and R-2 (ionization). (From Mulholland, G., Smoke Production and Properties, in *SFPE Handbook of Fire Protection Engineering*, 4th ed., National Fire Protection Association, Quincy, MA, 2008. Used with permission of Society of Fire Protection Engineers, copyright © 2008.)

response, or measured directly by mass concentration or particle number density. The reference measurement means are typically different than that utilized by the smoke detector being characterized. Therefore, the reliability of resulting data may vary.

Most smoke detector sensitivity testing is done using the principle of light attenuation (obscuration). The devices used are essentially beam-type detectors or optical density meters that measure an average reduction in light transmitted over the distance between the source and receiver. This distance is often on the order of a meter or more. Most common smoke detectors are fixed spot devices that measure smoke concentrations inside a sensing chamber. The amount of smoke in the sensing chamber may not be the same as what is outside the detector. Complicating this further is the fact that this optical density meter or smoke meter has its own response characteristics to smoke. For example, sooty black smoke may reduce light along an obscuration beam path, but not scatter much light in a photoelectric detector chamber. The relative response of the two sensors is affected by the characteristics of the smoke. In addition, the sensing mechanism may be different than that used for the detector. The amount of attenuation is sensitive to the type of light (e.g., color or polarization) and distance between the detectors for a given type of smoke (Mulholland, 2008).

Given a particular detector and burning material, all of these factors make calculating the activation of a specific detector very difficult. The complications increase significantly when more than one material is burning. Like all situations where the needed criteria are highly variable and difficult to determine, approximations using bounding criteria should be used. This also influences how the results of the analysis should be used.

Although a specific determination of detection criteria is difficult, a number of methods have been utilized (Schifiliti et al., 2008). These include using smoke concentration, a temperature rise analog to the smoke concentration, or a gas velocity analog to the smoke concentration. When using these various methods, the user should be aware of the limitations of each method to verify that it gives a bounding calculation for the specific situation encountered.

When measuring smoke concentration by light attenuation, a number of different quantities are reported. The intensity of light through the smoke is typically normalized by the intensity of light through clean air. This represents the fraction (or percentage) of light transmitted through the smoke. Most report the value as a percentage of attenuation known as percent obscuration (Schifiliti et al., 2008).

$$O = 100\left(1 - \frac{I}{I_0}\right)$$

where O is light obscuration (%), I is intensity of light received through smoke, and I_0 is intensity of light received through clear air.

The smoke concentration can also be reported as the log of the ratios of these two intensities, which is known as the optical density. It is a measure of the transmittance of light through smoke conditions for a given wavelength and distance. Optical density (Schifiliti et al., 2008) can be expressed as

$$D_\lambda = \log_{10}\frac{I_{\lambda,0}}{I_\lambda} = -\log_{10}\frac{I_\lambda}{I_{\lambda,0}} \tag{7.10}$$

where D_λ is optical density of smoke layer at wavelength λ, I_λ is intensity of light at wavelength λ received through smoke, and $I_{\lambda,0}$ is intensity of light at wavelength λ received through clear air.

The measured intensity of the light is a property of the instrument for which it was tested. Bouguer's law is used to normalize these values so that they are applicable to other situations (Mulholland, 2008). This law relates the intensity of light transmitted through smoke, the intensity of light transmitted through clear air, and the path length over which the light was transmitted. Technically, this law applies only to monochromatic light, but can be used to approximate the values for polychromatic light. The law, which can be written as base 10 or base e, is as follows:

$$\frac{I}{I_0} = 10^{-DL} = e^{-KL} \tag{7.11}$$

where L is distance over which the measurement was taken, D is optical density per meter, and K is light extinction coefficient (1/m).

A value of 2.303 ($D = K/2.303$) relates optical density per distance to the extinction coefficient.

Bouguer's law can also be used to normalize the amount of attenuation. When this is done, it is reported as the percent obscuration per unit distance (e.g., percent obscuration per meter). This value is

$$O_u = 100\left(1 - \left[\frac{I}{I_0}\right]^{1/L}\right) \tag{7.12}$$

where O_u is percent obscuration per unit distance.

The percent obscuration per meter is the value that is typically used for smoke detectors that are tested and approved. These are the values that are provided by the manufacturer or that can be programmed at the panel monitoring the smoke detection system. It specifies when a detector will activate for a specific smoke type and scenario. It may not correlate to when the detector will activate in a real-world fire.

The light extinction coefficient and optical density per unit distance are related to the smoke mass concentration through the use of a specific extinction coefficient or particle optical density, respectively. Seader and Einhorn (1977) determined that this relationship is approximately constant for wood- and plastic-type fuels. They found the particle optical density is 1.9 m²/g for smoldering fires and 3.3 m²/g for flaming fires, while the specific extinction coefficient is 4.4 m²/g for smoldering fires and 7.6 m²/g for flaming fires.

Mulholland and Croarkin (2000) found that the mean specific extinction coefficient, K_m, should be 8.7 m²/g + 1.1 m²/g for flaming fire smoke measured by red light. These ratios of optical density (or light extinction coefficient) to mass concentration can be used to estimate the mass concentration of smoke associated with a measured optical density value. For instance, an optical density of 0.050 m^{-1} is equal to a light extinction coefficient, K, of 0.115 m^{-1}. Using the specific extinction coefficient, K_m, from Mulholland and Croarkin, the smoke concentration can be estimated: 0.115/8.7 = 0.013 g/m^3.

Detector Sensitivity

Detectors sensitivities are reported to correspond with the measured light obscuration at the time of alarm during listing tests. When choosing a smoke concentration for smoke detection activation, one method is to use the reported detector sensitivity. Therefore, the detector is predicted to activate at that specific percent obscuration per meter for the smoke of the type used in the test. However, the actual sensitivity will vary depending on the fire conditions and the reference smoke meter used during the test. As previously discussed, the relationship between measured light obscuration and actual photoelectric or ionization detector response depends on the characteristics of the smoke present. Smoke from real fires may not have similar properties/relative responses to the smoke used in listing tests. Geiman and Gottuk (2003) evaluated using the detector sensitivity for a number of experimental fires. They found that most of the detectors required higher smoke concentrations to activate than the listed detector sensitivity and recommended that the listed detector sensitivity not be used.

Generic Sensitivity

One smoke concentration that is used to estimate smoke detector activation is an optical density per meter of 0.14 m^{-1} (NFPA, 2013). This value corresponds to an Underwriters Laboratories (UL) black smoke test requirement that is no longer conducted. According to research by Geiman and Gottuk (2003), this smoke concentration gives a fairly dependable prediction of detector operation: ionization detectors responded to 91% of flaming fires and 65% of smoldering fires; photoelectric (light scattering) detectors responded to 86% of flaming fires and 85% of smoldering fires at 0.14 OD/m. Based on these data, the 0.14 OD/m alarm threshold is a reasonable number to use, but is not bounding for all situations. It must be noted that detectors can activate at lower concentrations.

Temperature Analog

The temperature analog is another method that has been used for determining smoke detector activation. It was developed in the 1970s (Hesketad and Delichatios, 1977) as a simplified way to estimate smoke detector activation. It is based on the hypothesis that the smoke concentration (and optical density) is proportional to the temperature rise.

Table 7.4 Temperature Rise for Detector Response

Material	Ionization Temperature Rise		Light Scattering Temperature Rise	
	°C	°F	°C	°F
Wood	13.9	25	41.7	75
Cotton	1.7	3	27.8	50
Polyurethane	7.2	13	7.2	13
PVC	7.2	13	7.2	13
Average	7.8	14	21.1	38

Source: NFPA, *National Fire Alarm Code*, NFPA 72, National Fire Protection Association, Quincy, MA, 2013.

In order to be useful, this proportionality needs to be essentially constant. The original researchers believed, while the data showed some deviation, it was close enough to provide a reasonable approximation. Subsequent researchers (Geiman and Gottuk, 2003) have evaluated this and other data, and have questioned the constancy of this proportionality.

Geiman and Gottuk (2003) do not recommend using this method because the original assumptions are often not met. Others (Schifiliti et al., 2008) suggest that there are enough data and theory to show that a temperature analog is appropriate if the correct temperature analog is chosen. Like all of these methods, the temperature analog method should not be applied blindly. Using this method is appropriate if sufficient conservatism is used. In light of the controversy, more conservatism than normal may be warranted.

NFPA 72 (2013) recommends that the temperature analog be selected based on the type of fuel and detector, which are shown in Table 7.4. Schifiliti et al. (2008) give some recommended values and ranges of optical densities versus temperature that can be used to calculate temperature rises for various detectors.

The temperature analog should only be used in cases where there is a thorough understanding of the limitations associated with the method and how they are addressed.

Velocity Analog

All spot smoke detectors (ionization and photoelectric) require smoke to enter a chamber in order to be detected. This chamber creates a slight barrier to smoke entry. It was hypothesized that there may be a minimum velocity in order for smoke to enter the chamber and be detected. Brozovsky (1991) showed that 0.15 m/s (30 ft/min) was a critical velocity for at least some of the detectors tested. Alarms may still occur at lesser velocities, but the response may be delayed. Below this value, the optical density needed to activate the detectors rose considerably. Implicit in this value is the assumption that the smoke concentration in the 0.15 m/s airflow is sufficient to activate the detector.

Stratification and Other Issues

Smoke detection requires that smoke reach a detector sensing element. An analysis of the conditions within the space to determine conditions that could prevent smoke flow from reaching the detector must be made.

Typically, buoyancy is imparted to smoke by the heat of a fire and drives smoke flow upwards. Detection usually occurs early in the fire when the buoyancy is lowest (smallest fire size) and the smoke movement is most susceptible to other conditions. Smoldering fires burn with negligible heat release, making little buoyancy-driven flow. Fans and HVAC systems influence the flow of smoke, altering its natural, buoyant path. All analyses should address these possibilities.

Environmental conditions can also prevent smoke from reaching a detector. In areas where there is a temperature gradient with height, smoke may not reach ceiling-level detectors due to stratification. As smoke rises, it entrains fresh air and cools, which results in a smaller buoyant force. If the temperature of portions of a space is hotter than the plume, the buoyancy force can be reduced to zero, forming a layer of smoke that does not reach the ceiling. This layer will not rise to the detector until it grows to a sufficient size to overcome this temperature gradient. For example, at the roof of a tall glass-enclosed atrium during the summer, the temperature of the space may be hotter than the plume. Therefore, smoke will never reach the ceiling.

Since warm air rises due to buoyancy, temperature gradients may exist from mechanical heating, solar load, or lack of cooling. A smooth temperature gradient may exist from floor to ceiling, or there can be a discontinuity with a discrete step at a given height. The plume centerline temperature correlation can be used to determine the temperature expected of a smoke plume at a given height (Heskestad, 1984) and the maximum height of smoke rise (Heskestad, 1989):

$$\text{Discrete: } T_c = 25\dot{Q}_c^{2/3}z^{-5/3} + T_a \tag{7.13}$$

$$\text{Continuous: } Z_m = 5.54\dot{Q}_c^{1/4}\left[\frac{\Delta T_0}{dZ}\right]^{-3/8} \tag{7.14}$$

where ΔT_0 is rise in plume centerline temperature at detector height with respect to ambient temperature (°C), T_c is plume centerline temperature at height z (°C), T_a is ambient air temperature at fire level ($z = 0$) before ignition (°C), $\dot{Q}_c$ is convective portion of heat release (kW), z is detector height above fire (m), and Z_m is maximum height of smoke rise above the fire (m).

In both cases, the convective portion of the heat release rate is used as input. For the discrete case, the proposed height of the detector above the floor is used as input; the output is the expected temperature, T_c, at detector height. If is less than or nearly equal to the expected natural temperature near the detectors, the design must be ruled out or the desired size of fire to be detected must be increased.

For the continuous case, the expected temperature gradient is the input parameter, such as $\Delta T_0/dZ = 1.5°C/m$. The output is the maximum rise above the fire that smoke can be expected to travel. If the proposed design places detectors above or near this height, the design must be ruled out or the design fire must be enlarged.

Detection Calculation Methodologies

In the previous section, several different criteria for detector activation were introduced. This section will detail various methodologies for calculating if the detection criteria will occur.

Smoke Concentration (Optical Density Method)

One method for calculating the optical density is to assume that the smoke will be uniformly distributed throughout a volume. This volume can be the volume of the hot upper layer, which can be calculated using a zone model, or the entire volume of the space. The latter is only applicable if the smoke is uniformly distributed throughout the space, which may only be applicable for smoldering fires.

This method is detailed in NFPA 72 (2013) and the *SFPE Handbook* (Schifiliti et al., 2008). This method computes the optical density of resultant smoke based on burning characteristics of materials and the volume in question. Specifically, this quantity relates the amount of smoke expected from the mass burned. A detection time can be calculated by determining the total mass burned and the volume in which the smoke is dissipated as a function of time. This will usually require a numerical solution where the fire and its effects are discretized. This method relies on a variety of assumptions, including:

1. Smoke is uniformly dissipated in the volume of interest.
2. Gases are well mixed within the volume (uniform concentration).
3. Smoke can reach the ceiling.
4. Smoke can enter the detector.
5. There is negligible smoke deposition on surfaces.

The optical density of smoke in the space can be approximated by

$$D = \frac{D_m M}{V_c} \tag{7.15}$$

where D is optical density of smoke (m^{-1}), D_m is mass optical density (m^2/g), M is mass of fuel consumed (g), and V_c is volume in which smoke is dissipated (m^3).

The product of the particle optical density and the smoke yield for a material can be substituted for the mass optical density if appropriate data are not available.

A refinement on this method changes the definition of V_c to be the volume of a cylindrical region encompassing the jet on a flat ceiling. This refinement is detailed in NFPA 72 (2013). It uses the same assumptions as the previous method except that the volume is taken as the ceiling jet cylinder. Making the substitution

$$V_c = \pi r^2 h \tag{7.16}$$

an explicit relationship for distance can be obtained:

$$r = \left[\frac{D_m M}{D \pi h} \right]^{1/2} \tag{7.17}$$

where h is the ceiling jet depth, and r is the radial distance to the potential detector location.

Although it is often estimated that the ceiling jet thickness is approximately 10% of the floor-to-ceiling height of a space, this produces optimistic results. More cautious results are obtained by using a depth of 20 to 25%. However, this selection of a ceiling jet depth is arbitrary. It is also necessary to verify that the volume and radius that are calculated are physically real.

Models such as CFAST (Peacock et al. 2005), LAVENT (Davis and Cooper, 1989), and FDS (McGrattan et al., 2013) can be used to calculate smoke concentrations. They will be able to do this for flaming as well as smoldering fires provided the smoke production properties are known. FDS will be able to calculate smoke concentrations under most conditions, including HVAC airflow and buoyancy-limited smoke stratification. Zone models such as CFAST and LAVENT are unable to calculate these other effects. They typically use a two-layer assumption that is most applicable to flaming fires with high thermal buoyancy.

Temperature Analog

The main benefit of using a temperature analog as a detection criterion is the ease of calculating detector activation (Schifiliti et al., 2008). The same methodology as a heat detector is used. The only difference is that an artificial RTI (close to zero) is used. Typically this is selected to be 1.0 when the heat detector methodology is used for smoke detectors.

Velocity Analog

The main benefit of using a velocity analog as a detection criterion is the ease of calculating detector activation. The same correlations that are used for heat detector activation can be used. Caution should be used when applying some of the correlations as they rely on steady-state assumptions that may give unrealistic and nonconservative numbers because they do not account for transport times.

SUPPRESSION SYSTEM DESIGN

A variety of suppression systems might be used in a performance-based design. These include:

- Sprinkler
- Gaseous agent
- Foam
- Water mist
- Dry chemical
- Manual suppression

For all of these systems, it will first be necessary to detect the fire before the suppression system can be activated. Some of these suppression systems require a separate detection system (e.g., gaseous systems), while others do not (e.g., wet pipe sprinkler systems).

For many suppression systems, there is little information available to predict the effect that they have on a fire after they are activated. In such cases, an assumption is typically made—such as the fire is extinguished once the system is activated or the system maintains the fire at the same size it was when it was detected.

Sprinkler systems consist of a series of individual sprinklers that are connected to a water supply via piping. Most sprinklers are of the wet pipe type, where the piping is constantly charged with pressurized water. When a sprinkler is heated to its activation temperature, it opens and discharges water. Following the operation of the first sprinkler, subsequent sprinklers only open if they reach their activation temperature. The National Institute of Standards and Technology (NIST) developed a zeroeth-order approximation for determining the impact of sprinkler systems on fire size (Evans, 1993).

Dry systems are similar to wet systems, except that the piping is filled with pressurized air (or other gas) that holds a valve closed. Once a sprinkler is heated to its activation temperature, it opens and pressurized air is discharged. After the pressure in the piping is reduced below a certain amount, a valve opens and water enters the pipe. Deluge systems contain open sprinkler heads. The sprinkler valve is usually closed, and would open once a separate detection system is activated. Once the sprinkler valve is opened, water would flow from all sprinklers.

Preaction systems are similar to deluge systems, except the sprinklers are closed. In a preaction system, a separate detection system would open the sprinkler valve. However, water would not flow from any sprinkler until they were heated to their activation temperature.

Calculation of the time that it takes for the first sprinkler to activate has been possible for quite some time (see Calculations in Fixed Temperature Detectors). However, estimation of the operation of subsequent sprinklers or estimation of the effect of sprinkler activation on a fire has not been possible until the last few decades.

Evans (1993) developed a method to estimate the effect of sprinkler discharge on fire size based on research conducted by burning wood cribs (ordered arrays of wood sticks). Evans stated that this method would likely provide an upper bound for the effect of sprinkler activation on other fuels (meaning that suppression of fires involving fuels other than wood cribs would likely occur more quickly than or as quickly as fires involving wood cribs).

Evans was able to generalize the results by evaluating data from tests where sprinkler sprays were discharged on fires involving paper carts, desks, office mock-ups, office workstations, sofas, and wood cribs. Among these fuels, fires involving wood cribs were the most difficult to extinguish. Evans surmised that wood cribs were more difficult to extinguish than the other fuels because the fire is hidden from view within the array of sticks, while the other fuels burn primarily on their exposed surface.

In developing his method, Evans used data from fires in wood cribs that measured $610 \times 610 \times 610$ mm and $610 \times 610 \times 301$ mm (tall). Sprinkler discharge rates ranging from 0.026 to 0.133 mm/s were evaluated; however, some data at lower discharge densities had to be discarded because the fire was not suppressed at these low densities.

Evans developed the following correlation for the effect of sprinkler sprays on fires:

$$\dot{Q}(t - t_{act}) = \dot{Q}(t_{act}) \exp\left[\frac{-(t - t_{act})}{3.0 \dot{w}''^{-1.85}}\right]$$

where $\dot{Q}(t - t_{act})$ is heat release rate at time t following sprinkler activation (kW), $\dot{Q}(t_{act})$ is heat release rate at time of sprinkler activation (kW), t is time (s), t_{act} is time of sprinkler activation, and $\dot{w}''$ is water spray density (mm/s).

Fire Dynamics Simulator (McGrattan et al., 2013) models the effects of sprinklers on fire by estimating sprinkler activation, droplet trajectories, and tracking the water as it drips onto the burning commodity.

Unlike other thermal detector response prediction programs, Fire Dynamics Simulator is able to predict the response of sprinklers that activate after the first sprinkler. Operation of a sprinkler will disturb the fire plume and ceiling jet and will cool the fire environment. Additionally, water droplets can come into contact with unopened sprinklers. These effects are not considered by any models other than Fire Dynamics Simulator.

Fire Dynamics Simulator determines the initial size and velocity of water droplets based on user input values for the sprinkler that is modeled and the water pressure at the sprinkler. Droplet trajectories are calculated based on the initial sizes and velocities of droplets and their drag in the fire environment.

As a water droplet travels through air, it will lose mass as a function of the droplet equilibrium vapor mass fraction, the local gas phase vapor mass fraction, the heat transfer to the droplet, and the droplet's motion relative to the gas. This loss of mass will also absorb energy from the air. Fire Dynamics Simulator calculates this mass and energy transfer in a semiempirical manner. Similarly, water droplets will attenuate thermal radiation by scattering and absorption, and Fire Dynamics Simulator estimates these effects.

When a water droplet strikes a horizontal surface, Fire Dynamics Simulator assigns it a random horizontal direction, and the droplet moves at a fixed velocity until it reaches the edge of the surface, at which point it drops straight down at a fixed velocity. If the object is porous, a fraction of the water mass is assumed to be absorbed.

After accounting for the amount of water that actually strikes a burning surface, Fire Dynamics Simulator uses an exponential correlation similar to that developed by Evans (1993) to calculate the reduction in heat release rate. However, because Fire Dynamics Simulator is able to calculate the amount of water that reaches a burning surface, and Evans's method uses the total amount of water discharged by a sprinkler, the coefficients in Fire Dynamics Simulator are different than those developed by Evans.

SUMMARY

It is important to be able to estimate the activation of detection and suppression systems for a performance-based analysis. Heat detection is easier to predict than smoke detection. For smoke detection, there are many methods and associated shortcomings with each. The key is to understand and address the shortcomings as best as possible.

For suppression systems, once the activation time is determined, it is also necessary to determine the effect of the suppression system on the heat release rate.

REFERENCES

Alpert, R., Calculating Response Time of Ceiling-Mounted Fire Detectors, *Fire Technology*, 8(3), 181–195, 1972.

Alpert, R., Ceiling Jet Flows, in *SFPE Handbook of Fire Protection Engineering*, 4th ed., National Fire Protection Association, Quincy, MA, 2008.

Brozovsky, E., A Preliminary Approach to Siting Smoke Detectors Based on Design Fire Size and Detector Aerosol Entry Lag Time, Master's thesis, Worcester Polytechnic Institute, Worcester, MA, 1991.

Bryan, J., Behavioral Response to Fire and Smoke, in *SFPE Handbook of Fire Protection Engineering*, 4th ed., National Fire Protection Association, Quincy, MA, 2008.

Davis, W., *The Zone Fire Model Jet: A Model for the Prediction of Detector Activation and Gas Temperature in the Presence of a Smoke Layer*, NISTIR 6324, National Institute of Standards and Technology, Gaithersburg, MD, 1999.

Davis, W., and Cooper, L., *Estimating the Environment and the Response of Sprinkler Links in Compartment Fires with Draft Curtains and Fusible Link-Activated Ceiling Vents—Part II: User Guide for the Computer Code LAVENT*, NBSIR 89-4122, National Bureau of Standards, Gaithersburg, MD, 1989.

Evans, D., *Sprinkler Fire Suppression Algorithm for Hazard*, NISTIR 5254, National Institute of Standards and Technology, Gaithersburg, MD, 1993.

Evans, D., and Madrzykowski, D., *Characterizing the Thermal Response of Fusible-Link Sprinklers*, NBSIR 81-2329, National Bureau of Standards, Gaithersburg, MD, 1981.

Evans, D., and Stroup, D., *Methods to Calculate the Response Time of Heat and Smoke Detectors Installed below Large Unobstructed Ceilings*, NBSIR 85-3167, National Bureau of Standards, Gaithersburg, MD, 1985.

Geiman, J., and Gottuk, D., Alarm Thresholds for Smoke Detector Modeling, presented at Fire Safety Science—Proceedings of the Seventh International Symposium, International Association for Fire Safety Science, London, 2003.

Heskestad, G., Engineering Relations for Fire Plumes, *Fire Safety Journal*, 7, 25–32, 1984.

Heskestad, G., Note on Maximum Rise of Fire Plumes in Temperature-Stratified Ambients, *Fire Safety Journal*, 15, 271–276, 1989.

Heskestad, G., Fire Plumes, Flame Height, and Air Entrainment, in *SFPE Handbook of Fire Protection Engineering*, 4th ed., National Fire Protection Association, Quincy, MA, 2008, chap. 2-1.

Heskestad, G., and Delichatsios, M., *Environments of Fire Detectors—Phase 1: Effect of Fire Size, Ceiling Height and Material: Measurements*, vol. 1, NBS-GCR-77-86, National Bureau of Standards, Washington, DC, 1977.

Heskestad, G., and Bill, R., Quantification of Thermal Responsiveness of Automatic Sprinklers Including Conduction Effects, *Fire Safety Journal*, 14, 113, 1988.

Heskestad, G., and Smith, H., *Investigation of a New Sprinkler Sensitivity Approval Test: The Plunge Test*, FMRC Serial 22485, Factory Mutual Research Corporation, Norwood, MA, December 1976.

McGrattan, K., et al., *Fire Dynamics Simulator Technical Reference Guide*, NIST Special Publication 1018-6, National Institute of Standards and Technology, Gaithersburg, MD, 2013.

Mulholland, G., Smoke Production and Properties, in *SFPE Handbook of Fire Protection Engineering*, 4th ed., National Fire Protection Association, Quincy, MA, 2008.

Mulholland, G., and Croarkin, C., Specific Extinction Coefficient of Flame Generated Smoke, *Fire and Materials*, 24, 227–230, 2000.

Nam, S., Donovan, L., and Kim, J., Establishing Heat Detectors' Thermal Sensitivity Index through Bench-Scale Tests, *Fire Safety Journal*, 39, 191–215, 2004.

NFPA, *Engineering Guide to Performance-Based Fire Protection*, National Fire Protection Association, Quincy, MA, 2007.

NFPA, *National Fire Alarm Code*, NFPA 72, National Fire Protection Association, Quincy, MA, 2013.

Peacock, R., Jones, W., and Forney, G., *CFAST—Consolidated Model of Fire Growth and Smoke Transport (Version 5) User's Guide*, NIST Special Publication 1034, National Institute of Standards and Technology, Gaithersburg, MD, 2005.

Qualey, J., and Penney, S., Smoke Detector Response and Use of Antistatic Materials: An Empirical Study, *Fire Technology*, 41, 215–234, 2005.

Ruth, J., Odor Thresholds and Irritation Levels of Several Chemical Substances: A Review, *American Industrial Hygiene Association Journal*, 47, 142–151, 1986.

Schifiliti, R., Meacham, B., and Custer, R., Design of Detection Systems, in *SFPE Handbook of Fire Protection Engineering*, 4th ed., National Fire Protection Association, Quincy, MA, 2008.

Seader, J., and Einhorn, I., Some Physical, Chemical, Toxicological, and Physiological Aspects of Fire Smokes, presented at Proceedings of the 16th Symposium (International) on Combustion, The Combustion Institute, Pittsburgh, PA, 1977.

SFPE, *Evaluation of the Computer Fire Model DETACT-QS*, Society of Fire Protection Engineers, Bethesda, MD, 2002.

Walker, J., Hall, S., Walker, D., Kendal-Reed, M., Hood, A., and Niu, X., Human Odor Detectability: New Methodology Used to Determine Threshold and Variation, *Chemical Senses*, 28, 817–826, 2003.

Zalosh, R., *Industrial Fire Protection*, John Wiley & Sons, West Sussex, England, 2003.

Chapter 8

Smoke Control Design

INTRODUCTION

Smoke control can be provided by smoke barriers, natural vents, or mechanical systems. Design of the smoke control system takes into account a number of different methodologies and design considerations. Wind, stack effect, location of openings, buoyancy of fire gases, building ambient temperatures, and building tightness are all considered relative to their potential impact on smoke movement throughout the building.

There are many excellent resources on designing smoke control systems, so this chapter just provides an overview. For more information on designing smoke control systems, see Klote et al. (2012), Klote (2008), or Milke (2008).

STACK EFFECT

Stack effect is a phenomenon that will induce vertical airflow within a building due to the temperature differential between the building's inside temperature and the outside ambient air temperature. Generally, when the ambient air temperature is colder than the building temperature, air will move upward through the building. This is commonly referred to as normal stack effect and will occur under winter conditions.

When the ambient air temperature is warmer than the building temperature, air will move downward through the building. This is commonly referred to as reverse stack effect and will occur under summer conditions. The magnitude of the stack effect-induced airflow directly depends on the magnitude of the temperature differential between the interior and exterior of the building, as well as the building's height. Examples of the stack effect are depicted in Figure 8.1.

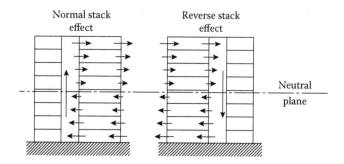

Figure 8.1 Stack effect flows in a building.

WIND

The effect of wind on a building is a very complex phenomenon, which is affected by building shape, building height, building construction, and even other nearby buildings. Wind pressures can vary widely across each face of a building, making input of wind pressure into computer simulations challenging. Wind is a transient phenomenon, and the magnitude of the wind and its direction can fluctuate. Therefore, many assumptions must be made in order to model wind in a smoke control analysis.

HVAC SYSTEMS

HVAC systems can impact airflows. The HVAC airflows typically have limited impact close to the fire where fire-generated flows will dominate. Farther from the fire, HVAC systems can be a primary driver. The impact of HVAC systems on smoke control systems will vary depending on whether dedicated systems are used exclusively for smoke control purposes, and if other building HVAC systems are designed to shut down upon actuation of the smoke control systems.

TEMPERATURE EFFECTS

For an un-sprinklered fire, buoyancy of hot fire gases can be a significant contributor to smoke movement through a building. Smoke will form a layer in the upper part of the fire compartment and adjacent spaces, and spread vertically via openings to floors above.

For a sprinklered fire, the contribution of these buoyant forces to overall smoke movement in the building is minimized. Once sprinklers are activated, the water spray will cool the hot gases. The momentum of the water spray will also stir the smoke layer, resulting in a more uniform smoke concentration within the room. In addition, activated sprinklers will suppress or control the fire, thereby reducing the smoke produced over time.

The smoke control systems for atriums may be based on the exhaust method of smoke control. Typically, sprinkler activation is not considered in these situations due to ceiling heights. For spaces adjacent to the atrium with low ceilings, it may be expected that the sprinklers will be effective at suppressing or controlling a fire.

In addition, atriums may have ambient temperature variations, such as may be caused by the heating under a glass dome. Therefore, a hot thermal layer may develop under the roof. In a tall atrium, the smoke cools as it rises. If the smoke has cooled, it may not penetrate the hot thermal layer.

SMOKE CONTROL METHODS

There are several different methods of smoke control, including the pressurization method, the airflow method, and the exhaust method, as design methods for mechanical smoke control systems. Passive smoke control systems are another option. The intent of smoke control is usually the provision of a tenable environment for the evacuation or relocation of occupants.

Pressurization Method

The pressurization method is intended to be used only when smoke barriers are present in a building. This method uses supply or exhaust ventilation to provide a minimum pressure differential across the openings in the barrier. A pressure differential of 25 Pa (0.05 in. of water) is commonly used in sprinklered buildings. This method is effective only when the proportion of the leakage area to the overall area of the smoke barrier is relatively small.

Airflow Method

For large openings where the pressurization method is not applicable, it is possible to prevent smoke from migrating through the openings by means of an opposed airflow to limit smoke migration from the fire zone through the opening. In general, it is also preferable that the area of the opening be relatively small in relation to the plane in which it is contained. In addition, the airflow method is typically only used for openings that are in the vertical plane.

Exhaust Method

The exhaust method of smoke control is typically utilized for large building openings connecting several floors, such as atriums and covered malls.

For smoke exhaust systems, the design objective is usually to maintain the accumulating smoke layer at least 2 to 3 m (6 to 10 ft) above any surface that forms a portion of a required egress system within a smoke zone.

Analysis of Smoke Filling

A common smoke control application is the protection of large open spaces such as atriums and covered malls. The primary application involves circumstances where these large open areas connect more than one level. The analysis of smoke control systems for these spaces is discussed in NFPA 92 (2012) and Milke (2008). The analysis must include elements such as fire location, fire size, atrium attributes (e.g., height, height of occupants in an atrium, openings, and volume), effects of sprinklers, and layer stratification.

Example

Develop a smoke exhaust rate for a 4-story atrium. The desired critical smoke layer interface height is 16.5 m. The design fire size is 5,000 kW. A balcony spill plume may also occur into the same atrium. The balcony from which the plume spills is located 7.8 m above the atrium floor. The width of the plume is limited by draft curtains to 5 m. Sprinklers under the balcony are predicted to activate at a fire size of approximately 2,000 kW.

AXISYMETRIC PLUME CALCULATION

Step 1: Calculation for convective portion of heat release rate.

$\dot{Q}_T$ = total fire heat release rate = 5,000 kW

$\dot{Q}_c$, the convective portion of the fire heat release rate, is assumed to be $0.7\dot{Q}_T$

$\dot{Q}_c = 0.7 \times 5000 = 3500 kW$

Step 2: Calculation for limiting elevation, z_1, compared to smoke layer height, z.

z_1 = limiting elevation corresponding approximately to the luminous flame height (m)

$z_1 = 0.166\dot{Q}_c^{2/5}$ (Equation 5.5.1.1d in NFPA (2012))

$z_1 = 0.166 \times 3500^{2/5} = 4.34$ m

z = critical design interface height above fire source = 16.5 m above atrium floor

Since $z > z_1$, the mass flow rate in the plume at height z is given by Equation 5.5.1.1e in NFPA (2012).

Step 3: Calculation for mass flow rate, $\dot{m}$, in plume at height, z.

$$\dot{m} = 0.071\dot{Q}_c^{1/3}z^{5/3} + 0.0018\dot{Q}_c \text{ (Equation 5.5.1.1e in NFPA (2012))}$$

$$\dot{m} = 0.071 \times 3500^{1/3} \times 16.5^{5/3} + 0.0018 \times 3500 = 121 \text{ kg/sec}$$

Step 4: Calculation for volumetric smoke exhaust rate, $\dot{V}$.

$$T_s = T_a + \frac{K_s\dot{Q}_c}{\dot{m}C_p} \text{ (Equation 5.5.5 in NFPA (2012))}$$

where T_s is smoke layer temperature (°C), T_a is ambient temperature (°C), K_s is fraction of convective heat release contained in smoke layer (–), and C_p is specific heat of plume gases (1.0 kJ/kg-°C).

As a conservative assumption (which will overestimate the smoke layer temperature), K_s is assumed to be 1.0. The ambient temperature is 20°C.

$$T_s = 20 + \frac{1 \times 3500}{121 \times 1.0} = 49°C$$

ρ_{smoke} = density of smoke (kg/m³)

$$\rho_{smoke} = \rho_a\frac{273 + T_a}{273 + T_s}$$

where ρ_a is density of ambient air = 1.2 kg/m³ at 20°C.

$$\rho_{smoke} = 1.2\frac{273 + 20}{273 + 49} = 1.09 \text{ kg/m}^3$$

$$\dot{V} = \frac{\dot{m}}{\rho_{smoke}} = \frac{121}{1.09} = 111 \text{ m}^3\text{/s} \approx 110\text{m}^3\text{/s} \quad \text{(Equation 5.7b in NFPA (2012))}$$

The volumetric exhaust rate, $\dot{V}$, required for maintaining smoke layer interface above critical design interface height, z (16.5 m above ground floor) is 110 m³/s.

SPILL PLUME CALCULATION

Step 1: Calculation for convective portion of heat release rate.

$\dot{Q}_T$ = total fire heat release rate = 2,000 kW

$\dot{Q}_c$, the convective portion of the fire heat release rate, is assumed to be $0.7\dot{Q}_T$.

$$\dot{Q}_c = (0.7) \times (2,000 \text{ kW}) = 1,400 \text{ kW}$$

Step 2: Calculation for mass flow rate, $\dot{m}$, in plume at height, z.

$$\dot{m} = 0.36(\dot{Q}_cW^2)^{\frac{1}{3}}(z_b + 0.25H) \quad \text{(Equation 5.5.2.1b in NFPA (2012))}$$

where W is width of balcony plume (m), z_b is height above underside of balcony to smoke layer interface (m), and H is height of balcony above base of fire.

$$\dot{m} = 0.36(1400 \times 5^2)^{\frac{1}{3}}(7.8 + 0.25 \times 8.7) = 117 \text{kg/s}$$

Step 3: Calculation for volumetric smoke exhaust rate, $\dot{V}$.

$$T_s = T_a + \frac{K_s\dot{Q}_c}{\dot{m}C_p} \quad \text{(Equation 5.5.5 in NFPA (2012))}$$

$$T_s = 20 + \frac{1 \times 1400}{117 \times 1.0} = 32°C$$

ρ_{smoke} = density of smoke (kg/m³)

$$\rho_{smoke} = \rho_a\frac{273 + T_a}{273 + T_s}$$

$$\rho_{smoke} = 1.2\frac{273 + 20}{273 + 32} = 1.15 \text{ kg/m}^3$$

$$\dot{V} = \frac{\dot{m}}{\rho_{smoke}} = \frac{117}{1.15} = 101 \text{ m}^3/s \approx 100 \text{ m}^3/s \quad \text{(Equation 5.7b in NFPA (2012))}$$

For spill widths up to and including 5 m, the previously calculated 110 m³/s exhaust quantity for the axisymmetric plume would be expected to be capable of exhausting smoke from a spill plume, as described by the above equation.

Note that this example does not include some elements such as stratification. The example is based on calculation of a defined layer height that correlates to the transition zone of the smoke layer (see Figure A.3.3.11.1 in NFPA (2012)). In an actual fire, some smoke will be present below this calculated height.

Analyses can also incorporate tenability of the exposure environment rather than just a smoke layer position. Typically in a CFD analysis, layer position must be defined using a fire exposure parameter, e.g., a certain visibility, temperature, or toxic environment.

REFERENCES

Klote, J., Smoke Control, in *SFPE Handbook of Fire Protection Engineering*, 4th ed., National Fire Protection Association, Quincy, MA, 2008, chap. 4-14.

Klote, J., Milke, J., Turnbull, P., Kashef, A., and Ferreira, M., *Handbook of Smoke Control Engineering*, ASHRAE, Atlanta, GA, 2012.

Milke, J., Smoke Management by Mechanical Exhaust or Natural Venting, in *SFPE Handbook of Fire Protection Engineering*, 4th ed., National Fire Protection Association, Quincy, MA, 2008, chap. 4-15.

NFPA, *Standard for Smoke Control Systems*, NFPA 92, National Fire Protection Association, Quincy, MA, 2012.

Chapter 9

Structural Fire Resistance

INTRODUCTION

Structures are designed to resist a set of loads. These loads consist of dead loads, live loads, and environmental loads. Dead loads are the loads imposed by the weight of the structure itself, including floors, walls, structural elements, and any other permanently incorporated items. Live loads include the loads associated with the use of the building, such as the equipment and furnishings placed in the building. Environmental loads include the loads placed upon the building by snow, wind, rain, earthquake, or flood.

Structural engineers consider these loads when designing building structures. The design must consider the structural properties of materials, such as steel or concrete, including yield strength, ultimate strength, and modulus of elasticity. Material properties used in designs are most often assessed at the ambient building temperatures. However, these properties change with temperature rise during a fire event.

Structural fire resistance is the ability of a structure, or portion thereof, to continue to support the imposed loads in the event of a fire. Fire affects the ability of a structure to support its load in several ways (Milke, 2008; Fleischmann et al., 2008):

- Reduction in yield strength as temperature increases
- Reduction in ultimate strength as temperature increases
- Reduction in modulus of elasticity as temperature increases
- Thermal expansion

Figure 9.1 shows the effect of elevated temperatures on A36 steel. Figure 9.2 shows how strengths of reinforcing steels used in concrete vary with elevation in temperature. Figure 9.3 shows how the compressive strength varies with temperature for concrete of various aggregates.

Unlike other material properties, the compressive strength of concrete can increase slightly with increase in temperature (see Figure 9.3); however, as the temperature continues to increase, the compressive strength

139

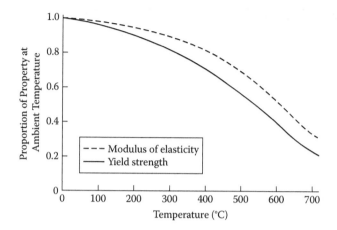

Figure 9.1 Properties of A36 steel at elevated temperatures. (From Milke, J., Analytical Methods for Determining Fire Resistance of Steel Members, in *SFPE Handbook of Fire Protection Engineering*, National Fire Protection Association, Quincy, MA, 2008. Used with permission of Society of Fire Protection Engineers, copyright © 2008.)

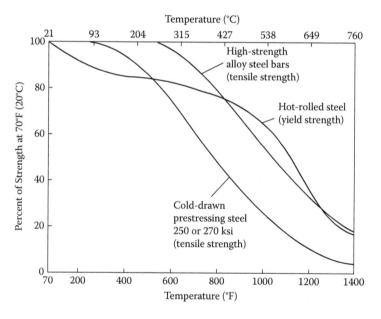

Figure 9.2 Properties of reinforcing steels at elevated temperatures. (From Fleischmann, C., Buchanan, A., and Chang, J., Analytical Methods for Determining Fire Resistance of Concrete Members, in *SFPE Handbook of Fire Protection Engineering*, National Fire Protection Association, Quincy, MA, 2008. Used with permission of Society of Fire Protection Engineers, copyright © 2008.)

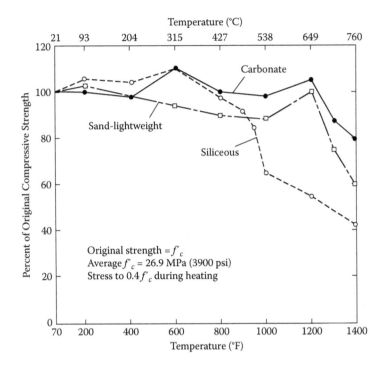

Figure 9.3 Properties of concrete at elevated temperatures. (From Fleischmann, C., Buchanan, A., and Chang, J., Analytical Methods for Determining Fire Resistance of Concrete Members, in *SFPE Handbook of Fire Protection Engineering*, National Fire Protection Association, Quincy, MA, 2008. Used with permission of Society of Fire Protection Engineers, copyright © 2008.)

eventually decreases. Concrete can also spall as it is heated, which can cause a reduction in concrete thickness. Increases in temperature also affect density, specific heat, and thermal conductivity.

Historically, structural fire resistance has been designed based on the standard fire resistance test. Individual structural assemblies, which consist of structural elements and any added protection, were subjected to a standard test, such as ASTM E-119 (ASTM, 2012). The standard test specifies the temperature in a test furnace as a function of time. Structural assemblies are assigned ratings based on their ability to support a load in the furnace or remain below peak point or average temperatures. The ratings are in units of time, which correspond to the length of time that the element is able to remain within the endpoint criteria specified by the standard test.

As an alternative to testing, empirically derived calculation methods can be used to determine the hourly rating of structural assemblies (ASCE, 2005).

Building codes specify minimum ratings for elements as a function of building characteristics such as use, height, and area. Application of the

traditional method of designing structural fire resistance has been simplified to the point that structural fire resistance is generally designed by
architects.

This method of designing structural fire resistance is based on research
conducted in the 1920s, and it has remained essentially unchanged (Ingberg,
1928). The underlying theory is that the product of temperature and time
integrated over the length of exposure provides a measure of fire severity.
Two different fire exposures with equal integrated time-temperature histo-
ries were considered to have equivalent fire severity.

For example, a compartment fire with a temperature of 1,200°C that
lasts 30 min would be assumed to have equal severity to a 900°C compartment fire that burned for 40 min. To avoid completely nonsensical comparisons, temperatures below a threshold of about 300°C would be neglected.

Minimum required fire resistance ratings in building codes were developed based on fire load surveys (which measured the mass of combustible
items per unit area) of typical occupancies and fire tests where the mass of
fuel per unit area was varied and the fire exposure (integrated product of
temperature and time) was measured.

There are a number of disadvantages to the traditional method of designing structural fire resistance:

- All of the factors that affect compartment fire temperature and duration are not considered. While the mass of combustibles is a factor,
  burning rate and temperature are also affected by the amount of ventilation available. Additionally, the thermal properties of the enclosure materials (density, specific heat, and thermal conductivity) will
  affect compartment fire temperatures.
- Single structural elements are tested in isolation, and structural performance at elevated temperatures is not considered.
- Heat transfer to bounding materials is not well represented by the
  product of temperature and time. While convection heat transfer can
  be represented by the product of time and temperature, radiation varies with temperature raised to the fourth power. A compartment fire
  with a temperature of 1,200°C would have approximately 2½ times
  the emissive power of a 900°C compartment fire.
- The fire load, whether expressed in units of mass per unit area (kg/
  m^2 or lb/ft^2) or energy potential per unit area (MJ/m^2), is not in and
  of itself representative of the fire hazard posed by combustible items.
  The ease with which a material burns is also a factor. For example,
  although wood has a heat of combustion that is approximately half
  that of most plastics (in other words, burning 1 kg of a plastic can
  liberate twice the energy of burning an equal mass of wood), wood's
  heat of gasification (a measure of how much energy it takes to create
  vapors) is two to five times that of most plastics. In ventilation-limited

fires, the rate of airflow into the enclosure will govern how much fuel vapor burns inside the enclosure, and fuel vapors that cannot burn inside the enclosure will vent from the compartment and burn outside. Therefore, a material that has a higher heat of combustion will not necessarily result in a more severe fire exposure within a compartment.

- The professionals that typically apply the traditional method of designing structural fire resistance generally do not have an understanding of fire behavior or structural performance.

Performance-based design of structural fire resistance can be done using the following approach:

- Determine the fire exposure to which a structure, or portion thereof, could be subjected.
- Determine the thermal response of the structure or portion thereof. This step involves conducting a heat transfer analysis based on the thermal boundary conditions determined in the previous step.
- Determine the structural response at elevated temperature. While a structural engineer would generally perform this task, fire protection engineers would provide input, such as material properties at elevated temperatures and strains induced by thermal expansion.

FIRE EXPOSURES

For most analyses of structural fire resistance, only fully developed fires are of interest. Heat transfer to the structure during fire growth typically provides only minimal heating in comparison to the heating during the fully developed stage. For insulated steel and concrete elements, consideration of the decay stage can be necessary as well, since heat will still transfer through the insulation, and the temperature of steel elements or steel reinforcement may continue to rise during the decay stage.

There are three types of fire exposures that are of interest for the analysis of structural fire resistance:

- Post-flashover compartment fires
- Local fire exposures
- Window flames

For most analyses, post-flashover fires are the exposure of interest. However, for compartments that are so large that a post-flashover exposure is not possible (e.g., arenas, airport terminals, etc.) or for exterior structures (e.g., bridges), consideration of local fire exposures will be required. For exterior structural elements that may be exposed to fires emanating from windows, it will be necessary to consider heat transfer from window plumes.

Generally, building codes permit some trade-off between structural fire resistance and sprinkler protection. For example, a building code may require columns to have a 3 h rating if sprinkler protection is not provided, and 2 h protection if sprinkler protection is provided. When designing structural fire resistance on a performance basis, sprinkler protection is generally not considered when determining fire exposures unless a risk-based design approach is used.

Sprinkler protection has the effect of reducing the frequency of fires that grow to a size that could affect a structure. This can be illustrated using the following equation:

$$F_{f,fire} = F_{fire}\left(P_{SSF}\middle|fire\right)\left(P_{fail}\middle|SSF\right) < F_{target}$$

where $F_{f,fire}$ is frequency of fire-induced failures (year^{-1}), F_{fire} is fire frequency (year^{-1}), $P_{SSF}|fire$ is given a fire, probability of the fire becoming structurally significant (–), $P_{fail}|SSF$ is probability of structural failure given a structurally significant fire (–), and F_{target} is target failure frequency (year^{-1}).

The fire frequency ($F_{f,fire}$) is the frequency that fires, of any size, would be expected to occur in the building. Not all fires grow to a size that they could affect the structure. For example, a fire that starts in a trash can that is quickly extinguished by occupants would not impact the ability of the building structure to continue to support loads.

Only fires that grow to a size that they become structurally significant affect the ability of the building structure to support loads. The fraction of fires that grow to become structurally significant is represented by the term $P_{SSF}|fire$. It is in this term where the use of fire protection systems, like sprinkler systems or fire alarm systems, is considered. Use of a fire protection system in the building would reduce the $P_{SSF}|fire$.

Table 9.1 provides $P_{SSF}|fire$ for offices for buildings of a variety of construction methods. As can be seen in Table 9.1, $P_{SSF}|fire$ is lower when sprinklers or a fire alarm system, or both, are installed.

A fire that is structurally significant will not necessarily cause failure of the structure. This is represented by the term $P_{fail}|SSF$, which represents the fraction of structurally significant fires that cause structural failure. The use of structural fire resistance materials, such as spray-applied materials or encasement, would reduce this term.

The acceptance criteria would be expressed in terms of frequency of structural failure. A typical value would likely be on the order of 10^{-6} per year, which means that there would be a 1/1,000,000 chance of fire-induced structural failure in any year.

Table 9.1 Fraction of Structurally Significant Fires in Offices

Construction	No Sprinklers or Alarm	Sprinklers with No Alarm	No Sprinklers with Alarm	Sprinklers and Alarm
Fire resistive	0.13	0.04	0.07	0.03
Protected noncombustible	0.15	0.05	0.06	0.03
Unprotected noncombustible	0.19	0.07	0.10	0.05
Protected ordinary	0.21	0.03	0.10	0.04
Unprotected ordinary	0.30	0.11	0.17	0.07
Protected wood frame	0.30	0.13	0.18	0.08
Unprotected wood frame	0.37	0.12	0.20	0.07

Source: NFPA, *Standard for Determination of Fire Loads for Use in Structural Fire Protection Design*, NFPA 557, National Fire Protection Association, Quincy, MA, 2012.

THERMAL RESPONSE

Once the fire boundary conditions have been estimated, the thermal response of the structure can be determined. Because the problem is generally nonlinear, a discretized numerical solution is often required.

Determining the thermal response of a structure is complicated by the fact that thermal properties vary with temperature. For design purposes, single values of these properties can be used if they would yield conservative estimates. In cases where a thermal conductivity or specific heat increases with temperature, ambient temperature values can be used (since this simplification would result in a greater estimate of temperature rise than using temperature-dependent properties).

Relatively simple numerical solutions are available for estimating the response of columns. However, computer-based finite element or finite difference solutions are generally required for beams because the slabs act as heat sinks. For concrete, computer analysis is generally required. Computer-based heat transfer programs include FIRES-T3 (Iding et al., 1977), TASEF-2 (Paulsson, 1983), SAFIR (Franssen et al., 2001), and SUPER-TEMPCALC (Anderberg, 1985).

DESIGN OF STRUCTURAL FIRE RESISTANCE

In some cases, protection will have to be provided to a structure, or portions thereof, to increase the probability that temperatures remain below those where unacceptable structural performance would result. For steel elements, insulation can be provided. The following methods of protection can be used (Milke, 2008):

- Board products, such as gypsum board, fiber-reinforced calcium silicate board, vermiculite-sodium silicate board, and mineral fiber board.
- Spray-applied materials, such as cementitious plasters, mineral fibers, magnesium oxychloride cements, and intumescents.
- Concrete encasement.
- Membranes, such as suspended ceilings.
- Flame shields, which are used to protect portions of a structure from flame impingement. Flame shields are generally only used in designs where exposure to a post-flashover is not possible, i.e., elements that may be exposed to localized fires or window flames.
- Heat sinks, where a hollow element is filled with water or concrete.

Structural fire resistance is typically provided for concrete by providing additional concrete cover to reinforcing elements. Additional concrete cover is typically needed because moving reinforcing elements further from the exposed surfaces of the concrete would reduce the moment resisting capability these elements provide.

Structural engineers usually consider various load combinations for design. This is done in recognition that structural loads are stochastic, and it is unlikely that the worst case loads would all occur simultaneously. For example, it is highly unlikely that the worst case snow would occur simultaneously with the worst case wind and an earthquake. These load combinations are based on a design failure rate.

NFPA 5000 (2012b) provides the following load combination for fire conditions:

$$1.2D + T + 0.5L + (0.5L_r \text{ or } 0.2S)$$

where D is dead load, T is structural action resulting from thermal expansion, L is live load, L_r is live roof load, and S is snow load.

This load combination would apply to high-rise buildings (23 m (75 ft) or taller) of importance groups III or IV. These importance groups include buildings that represent a substantial hazard to life in the event of failure and buildings that provide essential community services.

REFERENCES

Anderberg, A., *PC-TEMPCALC*, Institutet for Brandtekniska, Fragor, Sweden, 1985.

ASCE, *Standard Calculation Methods for Structural Fire Protection*, ASCE-SFPE 29-05, American Society of Civil Engineers, Reston, VA, 2005.

ASTM, *Standard Test Methods for Fire Tests of Building Construction and Materials*, ASTM E-119, American Society of Testing and Materials, West Conshohocken, PA, 2012.

Fleischmann, C., Buchanan, A., and Chang, J., Analytical Methods for Determining Fire Resistance of Concrete Members, in *SFPE Handbook of Fire Protection Engineering*, National Fire Protection Association, Quincy, MA, 2008.

Franssen, J.M., Kodur, V.K.R., and Mason, J., *User's Manual for Safir 2001: A Computer Program for Analysis of Structures Submitted to the Fire*, University of Liege, Belgium, 2001.

Iding, R.H., Nizamuddin, A., and Bresler, B., *UCB FRD 77-15*, University of California, Berkeley, 1977.

Ingberg, S., Tests of the Severity of Building Fires, *Quarterly of the National Fire Protection Association*, July 1928, pp. 43–60.

Milke, J., Analytical Methods for Determining Fire Resistance of Steel Members, in *SFPE Handbook of Fire Protection Engineering*, National Fire Protection Association, Quincy, MA, 2008.

NFPA, *Standard for Determination of Fire Loads for Use in Structural Fire Protection Design*, NFPA 557, National Fire Protection Association, Quincy, MA, 2012a.

NFPA, *Building Construction and Safety Code*, NFPA 5000, National Fire Protection Association, Quincy, MA, 2012b.

Paulsson, M., *TASEF-2*, Lund Institute of Technology, Sweden, 1983.

Chapter 10

Fire Testing

INTRODUCTION

Fire testing was originally established to provide a method to classify the fire resistance, flame spread, or other characteristic of an assembly or material. Typical data provided were limited to pass/fail, a rating, or an arbitrary classification. Fire protection engineering has developed to allow the numerical analysis of fire safety issues, and fire testing has evolved to address the need for information. In the last 20 years, standardized testing has evolved from test methods that provide limited insight to actual fire performance to test methods that provide direct input for performance-based design applications.

This chapter provides an overview of how fire testing is related to and implemented in performance-based design. A trend toward the use of performance-based design has increased the demand for data that describe actual performance in end use conditions. Fire growth models such as the zone model Consolidated Fire and Smoke Transport (CFAST) and the computational fluid dynamics (CFD) model Fire Dynamics Simulator (FDS) need suitable fire properties for the input materials.

Many times, standard test methods have to be modified or nonstandard testing has to be conducted to fill a void and provide the data needed for engineering calculations. This process has expanded the role of fire testing and forced the standard test methods to become more flexible.

Fire testing is growing due to its importance as a source of input data and also in response to the development of new products. Testing provides information such as ignition time, fire growth rate, heat release rate, smoke production rate, toxicity, extinguishment times, fire spread, smoke spread, and the fire's impact on the structure as a whole. In many cases, testing is not just conducted on a single material sample, but on complete assemblies, providing performance results for real-world applications. The overall objectives of testing may include:

- Developing data for models and fire safety engineering calculations

- Establishing the adequacy of fire safety systems and their design parameters for new types of systems or for existing systems used in new scenarios
- Evaluating the hazards associated with materials or configurations

The information obtained can be used to enhance the requirements of the prescriptive codes, complete performance-based analysis, assist in fire investigations or product development, or perform fire model validation. This increase in fundamental knowledge is needed for advancing fire safety engineering, and with it, the safety of products, facilities, transportation, first responders, and people.

FIRE TESTING

Demonstrating adequate performance to the pass/fail criteria established in prescriptive building codes is a necessity for a material or assembly to be implemented in a prescriptive design. The performance criteria are typically stated in terms of passing a standardized test, which may or may not reflect real-world fire exposures. For example, ASTM E 84 (2013a) evaluates the flame spread and smoke production of an interior finish material exposed to a standardized fire exposure in a specific orientation. This test configuration does not replicate all exposures, yet almost all interior finish materials are prescriptively regulated by the building codes based on the flame spread and smoke production criteria established from this standardized test.

The limitations of the ASTM E 84 criteria were recognized for textile wall covering materials. After review of a number of large fires involving Class A (the highest classification available in the standard) wall covering materials tested in accordance with ASTM E 84, it was found that the standard exposure did not reflect the actual hazard potentially posed by the material. Comparative room corner tests were conducted using Class A materials in which flashover occurred. Based on the results, the standard ASTM E 84 test method was modified and alternative test methods for specific materials were developed (Belles et al., 1987).

The standard test methods are evolving to permit testing of materials to allow application to performance-based design. Some of these test methods are discussed below.

Test Standards and Guides

The American Society for Testing and Materials (ASTM), Underwriters Laboratories (UL), National Fire Protection Association (NFPA), and Factory Mutual (FM) have produced standard test methods and guides by which the fire performance of systems, assemblies, and materials can be evaluated. Some test methods have application, however limited it may be, to performance-based design. Some common test methods and their applicability to performance-based design are discussed, along with a brief summary of the test, test apparatus, and test standards. For some tests, UL, NFPA, and FM have essentially identical versions of the same ASTM test method. Other originations, such as the International Organization for Standardization (ISO), have similar tests as well. In the sections that follow, the ASTM versions of similar test methods are discussed.

ASTM E 119 (2012a): Standard Test Methods for Fire Tests of Building Construction and Materials

Knowing the performance of walls, columns, floors, and other building members when exposed to fire conditions is of major importance in fire-resistant construction. ASTM E 119 utilizes a standard time-temperature curve that describes an exposing fire of controlled severity and extent. The performance of the tested assembly is defined as the period of time elapsing from ignition to the first sign of a critical point behavior. This provides an hourly rating of the building member.

Typical test results are provided solely as an hourly rating. Therefore, this test has limited use in a performance-based design without specific details of the test results, including actual times, temperatures, and failure mechanisms. With detailed results, one may evaluate how variations in actual end use construction from listed assemblies would affect the anticipated rating of the assembly. Detailed test results are typically only provided to the manufacturer that paid for the tests. The manufacturer may or may not release this information. If available, these data are useful for validating heat transfer models that might be used in a structural fire resistance design or to estimate thermal properties of fire-resistant materials.

This standard can also be used to establish endpoint criteria for failure in structural analyses. For example, failure temperatures for beams and columns are established. In a performance-based design analysis, a calculation of steel temperature based on these failure points can be used. However, the failure points used in the standard may or may not be indicative of fire performance in the end use condition.

The tests can also be modified by changing any number of variables to fit the needs of a performance-based design. For example, the exposure fire severity can be adjusted and the member can be tested to see if it meets the objectives of the performance design. However, full-scale standardized tests can be expensive. Therefore, computer modeling of exposures or nonstandard smaller-scale tests similar to ASTM E 119 can be utilized.

ASTM E 1529 (2013e): Standard Test Methods for Determining Effects of Large Hydrocarbon Pool Fires on Structural Members and Assemblies

ASTM E 1529 (2013e) evaluates the performance of a structural steel element exposed to a flammable or combustible liquid fire to determine the fire resistance rating of the protection material applied to the steel element. The ASTM E 1529 exposure temperatures increase much faster than those in ASTM E 119; therefore, in a performance-based design where a design fire scenario involves extensive exposure to a flammable liquid fire, the results from ASTM E 1529 might be more relevant.

Another example is to consider the fire resistance rating of an insulated duct. For this example, a flat 3×3 m (10×10 ft) wall sample is constructed and subjected to the ASTM E 119 time/temperature exposure criteria. The test specimen consists of ceramic fiber insulation installed on the exposed face of 1.6 mm thick (16 gauge) sheet steel. Steel studs are attached to the unexposed face of the wall assembly to provide structural support. Using the test results (unexposed surface temperatures as a function of time), and generally knowing the thermal properties of ceramic fiber blanket, a thermal heat transfer model can be used to model the performance of the test specimen subjected to the ASTM E 119 fire exposure to refine the thermal parameters for the insulation material. With this validated model for the insulation, the insulation thickness, insulation orientation, steel thickness, and exposure criteria can be adjusted to evaluate the change in performance of the insulated duct.

ASTM E 2257 (2013g): Standard Test Method for Room Fire Test of Wall and Ceiling Materials and Assemblies

This test standard is meant to evaluate the contribution to room fire growth provided by wall or ceiling materials under specified fire exposure conditions. The material is installed in a manner representative of field use on the walls or ceiling and exposed to a standard fire exposure source representing a small, growing fire source. The test is conducted in a $2.4 \times 3.7 \times 2.4$ m (tall) ($8 \times 12 \times 8$ ft) room containing a single open doorway.

Acceptable performance is determined by the amount of flame spread across the surface of the test sample, smoke production, and whether flashover occurs. As is the case with ASTM E 119, any of the assumptions or variables can be adjusted, such as the fire exposure condition, to suit the needs of a performance-based design.

This test provides a realistic indication of how wall linings will perform in a performance-based design. Application of the test results to room geometries different from the test room can be accomplished by using a fire model. From the test data, the smoke yield produced during a design fire can be calculated.

Inputting the calculated smoke yield and the measured heat release rate, the impact on the development of untenable conditions from the textile wall covering material in different room geometries can be predicted.

ASTM E 2067 (2012b): Standard Practice for Full-Scale Oxygen Consumption Calorimetry Fire Tests

This standard provides methods to construct, calibrate, and use large-scale oxygen consumption calorimeters to help minimize testing result discrepancies between laboratories. The methodology described in the standard is used in a number of ASTM test methods, in a variety of nonstandardized test methods, and for general research purposes.

The practice does not provide a pass/fail criterion, nor does it describe a test method for any material or product. Standard practices such as this one describe the proper methodology for conducting oxygen consumption calorimetry testing. The data generated from these types of tests lend themselves to performance-based analysis by being applicable to a wide range of test setups and nonstandardized test methods. Figure 10.1 shows the oxygen consumption calorimetry hood.

The following example illustrates how this test method might be used in a performance-based design application. The heat release rate, smoke production, and radiant heat flux of a typical office configuration may be required for conducting a life safety analysis. In order to generate the

Figure 10.1 Oxygen consumption calorimetry. (Image courtesy of Jensen Hughes.)

required test input data, a mock-up of an office configuration is constructed and placed onto a load cell under a heat release rate calorimeter. Heat flux transducers are located at various distances from the test sample to record the heat flux output. A small ignition source is used to initiate the fire, and all products of combustion are captured in the heat release rate calorimeter. The test results can be input into a computer model to predict the smoke layer development rate, rate of descent, and temperature to predict the time when occupant egress would not be possible.

ASTM E 603 (2013b): Standard Guide for Room Fire Experiments

ASTM E 603 is written to assist those planning to conduct full-scale compartment fire experiments. Room fires are conducted to provide information on the performance of materials subjected to a standard fire exposure source to meet prescriptive code requirements. These tests also generate data that can be used as input for fire modeling or the validation of fire models.

This standard guide provides planning assistance. This allows the user the ability to reference this standard along with assumptions and variables that can be accepted by an authority having jurisdiction (AHJ) more readily than if nonstandardized testing had been used. Figure 10.2 shows a room fire test experiment.

ASTM E 1623 (2011): Standard Test Method for Determination of Fire and Thermal Parameters of Materials, Products, and Systems Using an Intermediate Scale Calorimeter (ICAL)

This standard test method evaluates the propensity for ignition of an exterior wall assembly exposed to a radiant heat source for a specified duration.

Figure 10.2 Room fire test. (Image courtesy of Jensen Hughes.)

The method is limited to the testing of planar, or nearly planar, exterior wall assemblies. Fire endurance, surface burning characteristics, and the influence of openings on the propensity for ignition are not evaluated by the test method. While the standard test method provides limited data for a performance-based design, evaluation at different heat flux levels can provide a relationship between heat flux and ignition propensity.

The ICAL is typically used to evaluate the performance of an exterior insulation finish system (EIFS) exposed to a constant heat flux for a specified duration. In a performance-based design, the ignition characteristics of the EIFS system are required to be known for a range of heat flux levels. The heat flux can be changed by adjusting the separation distance between the EIFS system and the radiant heat source, and the performance of the test sample can be evaluated. Using the test data, the performance of an EIFS system installed on the exterior portion of a building near a low roof can be evaluated for various roof fire exposures.

ASTM E 2058 (2013f): Standard Test Methods for Measurement of Synthetic Polymer Material Flammability Using a Fire Propagation Apparatus (FPA)

The FPA was originally developed by Factory Mutual (now FM Global) to quantify the flammability characteristics of polymeric materials used in clean room applications. Test data generated using the FPA include time to ignition, effective and convective heat release rates, mass loss rates, and smoke extinction coefficients. In this test, a $100\ mm^2$ (4 in.^{2}) sample is exposed to an incident radiant heat flux of between 0 and 65 kW/m^2. The products of combustion are collected in a hood where gas analysis and smoke production measurements are performed. Test data describing the transient response of a material to a prescribed heat flux are generated and can be used as an input into computer models.

The following example illustrates how to use results from ASTM E 2058 and other small-scale ignition tests. A flame spread model is used to predict the lateral flame spread up a vertical surface when exposed to a small initiating fire source. Small-scale testing of representative samples in the cone calorimeter (ASTM E 1354), the fire propagation apparatus (ASTM E 2058), or the lateral ignition and flame transport (LIFT) apparatus (ASTM E 1321) provides model input data such as heat release rate, mass loss rate, heat of combustion, and smoke production. The performance of a material under various ignition sources can be predicted using these data to evaluate the time to the onset of hazardous conditions.

ASTM E 1354 (2013d): Standard Test Method for Heat and Visible Smoke Release Rates for Materials and Products Using an Oxygen Consumption Calorimeter

The cone calorimeter is used to measure properties such as surface ignition, fuel mass loss, and heat release rates. Cone calorimeter testing is conducted on solid materials, such as wood samples or composites, in a similar manner as the FPA (ASTM E 2058). The sample (100×100 mm (4×4 in.)) is placed in a steel holder and exposed to a conical-shaped heating element designed to provide a uniform heat flux across the surface. Figure 10.3 shows a schematic view of the cone calorimeter.

The time to ignition, duration of burning, and mass loss is recorded. The smoke generated by the sample is measured, and the heat release rate can be calculated by oxygen consumption calorimetry. The cone calorimeter can

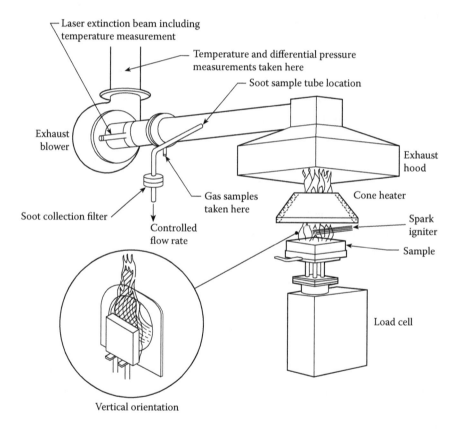

Figure 10.3 Schematic view of the cone calorimeter. (From Babrauskas, V., The Cone Calorimeter, in *SFPE Handbook of Fire Protection Engineering*, National Fire Protection Association, Quincy, MA, 2008; used with permission of Society of Fire Protection Engineers, copyright © 2008.)

also be used to determine the critical radiant heat flux necessary for ignition for a particular material. The cone calorimeter provides material fire properties without the costs associated with large-scale testing. These fire properties can be used to model the product in a large-scale application.

Most of the parameters obtained from the cone calorimeter (and the FPA) can be utilized as inputs to fire modeling. For example, an engineer may want to evaluate the hazard of plastic containers stored in a warehouse. A cone calorimeter test will provide heat release rate data that can be used in modeling for the hazard to the structure.

An example of using the ASTM E 1354 test method is as follows:

- An engineer has been requested to evaluate the structural fire hazard posed by plastic sheets stored in a warehouse. The concern is that the structure is large and the owner wants to build using unprotected construction, even though the code mandates protected construction. The AHJ wants to know if the storage/design fire can potentially cause structural failure.
- Heat release rate data are not available for the plastic material. Therefore, test data are required to develop design fire curves.
- Small-scale samples are tested in the cone calorimeter to measure the burning characteristics. Plastic burns in place and has a specific heat release rate per unit area. Full-scale tests are performed to establish flame spread and fire growth characteristics. Heat release rate is then applied over an exposed area to determine design fire curve.

ASTM E 1678 (2010): Standard Test Method for Measuring Smoke Toxicity for Use in Fire Hazard Analysis

The pyrolysis or combustion of every combustible material or product produces smoke that can be toxic. The E 1678 test method provides a means for determining the lethal toxic potency of smoke produced from a material or product ignited while exposed to an incident heat flux of 50 kW/m^2. A 75×125 mm (3×5 in.) sample is placed in a combustion chamber and exposed to a constant heat flux for 30 min. The combustion products pass through an exposure chamber. Either laboratory mice or combustion gas analysis is used to assess the toxic potency of the material being tested. The data generated in the E 1678 tests can be used as inputs in a fractional effective dose (FED) analysis to assess the toxicity of a material to building occupants.

For example, if a new product is being assessed for approval, and the issue of toxicity is required to be addressed, an FED analysis could be conducted to predict the toxicity hazard of the material when involved in a fire. A sample of the material is tested in accordance with either ASTM E 1678 or E 800 (see below) to determine the concentrations of CO, HCl, and HCN. Once the small-scale gas concentrations are determined, the

FED analysis can be used to predict the lethal concentrations of toxic gases present in a building for a simulated fire.

ASTM E 800 (2007): Standard Guide for Measurement of Gases Present or Generated during Fires

ASTM E 800 provides guidance on conducting measurement of combustion gases (CO), CO_2, O_2, NO_X (nitrogen oxides), SO_X (sulfur oxides), and other elements. The standard provides several techniques for measuring each gaseous species. The potential advantages and disadvantages of each measurement method are also discussed. This is a standard guide similar to ASTM E 603, providing best practices and methods for measurements, making it applicable to a number of potential test setups.

ASTM E 1321 (2013c): Standard Test Method for Determining Material Ignition and Flame Spread Properties

The lateral ignition and flame spread test, commonly referred to as LIFT, is a bench-scale test that provides flame spread data for samples exposed to an incident radiant heat flux. The test results identify a minimum heat flux and temperature necessary for ignition, lateral flame spread rate, effective thermal inertia data, and a flame heating parameter relative to flame spread.

LIFT data can be input into computer models to predict a material's performance. The sample tested is 760 mm long (30 in.) by 125 mm (5 in.) wide. While not full scale, the data generated from a material tested in the LIFT can be used as inputs to full-scale models in a performance-based design.

Approval of Systems

Underwriters Laboratories (UL) and FM Global (FM) also provide testing for various fire safety systems, such as fire alarm and suppression systems. When a new agent or type of system is developed, such as water mist, FM-200, Intergen, or video detection, UL and FM may evaluate the specific manufacturer's system fire performance characteristics.

Whether performance is measured by detecting or extinguishing a fire, the system is evaluated to establish design parameters and capabilities. Typically, standardized test methods are employed to evaluate the performance of the product. Therefore, the amount of usable data for a performance-based design may be limited. However, in many cases, the standardized tests were developed based on performance testing of the systems. This performance testing can provide insight into when a system should function as anticipated.

Fire testing is utilized in the development of suppression system design requirements in order to evaluate system performance in a real fire situation. Typically, a fire scenario is developed that represents a specific fire hazard or a generic class of fire hazards. The suppression system is tested against the developed scenario, with its design parameters varied until the minimum design requirements for successful suppression of the fire scenario have been established. These requirements are increased by a factor of safety and form the basis of the system design for actual installations involving the hazards represented by the developed fire scenario. The established basis of system design is then generally manifested in the system's listing and approval documentation.

For gaseous agents, minimum agent design concentrations, minimum average nozzle pressures, and maximum nozzle spacing are established in a series of tests involving fire scenarios representing generic Class A and Class B fire hazards. In a UL Listing and a FM Global Approval Test program, the Class A fire scenarios include elevated wood crib fires and a series of baffled plastic sheet array fires. The Class B fire scenarios include a large elevated heptane pan fire and a scenario involving 10 small heptane pans distributed around the test enclosure.

The International Maritime Organization (IMO) adds fire scenarios more representative of the fire hazards associated with shipboard machinery spaces, including diesel and heptane spray and pan fires. Specific tests are often required to evaluate performance in fires involving materials not represented well by these generic hazard scenarios. The protection of exotic materials and configurations, such as anechoic chambers, generally would require specific additional tests.

For water mist systems, the minimum application rate depends on the specific fire hazards and configurations, as opposed to just the design concentrations for a gaseous agent. Therefore, the listings and approvals are more narrowly defined and the scenarios used for fire testing are more specific to a given hazard. For example, a water mist listing may cover only a single type of enclosure within a specified volume range.

In situ testing of fire safety systems installed in buildings, sometimes referred to as a commissioning test, provides another method of testing for approval of systems. Smoke control systems are the most common systems tested in a building through actual fire exposure or fire simulation. To test a smoke control system, heated smoke is required for the smoke to rise to the location of the exhaust fans, which are typically at the top of the atrium. Figure 10.4 shows an *in situ* smoke control system test.

A heat generation source is required to provide the necessary smoke buoyancy. In much of the testing done where actual smoke is used, a gas burner or other fire source is used to create a hot smoke layer to test specific capabilities of a system.

Figure 10.4 Smoke test in atrium. (Image courtesy of Jensen Hughes.)

Nonstandard Testing

Nonstandard testing is testing performed that may not explicitly fit into the guidelines established by the published standards or guides. The tests may include elements that are measured by ASTM standards, such as a cone calorimeter, but require engineering judgments beyond following a strictly defined approach.

Nonstandard testing has been increasing due to widespread implementation of performance-based design. Nonstandard testing expands the knowledge base and answers questions posed as different situations and configurations are encountered.

With the increased demand for human behavior research, testing that is anything but standard is conducted in laboratories and real-world locations to help explain patterns and choices made by occupants. Data can even be obtained from video or eyewitness accounts of real fire events, such as testimonials from World Trade Center evacuees (Averill et al., 2005).

These data have explained the mindset and mentality of escapees and can be used to better design signaling and escape routes in future buildings using performance-based codes.

Nonstandard testing may also be utilized to develop protection schemes for unique occupancies or uses. An example is as follows:

- An engineer has been requested to determine the required attributes of a new method of fire detection and suppression in ordnance magazines. Because of the high risk and unique features (weapons automation), it was determined that a performance-based design was required.
- The design approach considers the best combination of fire detection and suppression systems in a performance-based design to avoid weapons detonation (referred to as cook-off) and inadvertent activation of the systems.
- Testing could be conducted in a mock magazine to determine the fire size and time needed to reach ordnance cook-off. In addition, various detection systems could be implemented and the activation times recorded during a variety of fire and nuisance scenarios. Tests could be conducted on various suppression systems and their extinguishing capabilities and timelines to establish when detection, activation, and suppression would be required to prevent cook-off. The impacts of inadvertent activation of the system could also be evaluated for each of the alternatives.

TESTING OUTLINE AND PLAN

Test results must be provided in a manner that is coherent and thorough to be useful for modeling or design. It is important to present this information in a format that is easily understood by people who are not engineers. The following is an example outline for a test plan or test report with brief descriptions on each section of the document.

Introduction

The introduction establishes the background information and provides the reader with the necessary knowledge to understand the document. This usually contains the driving force behind the testing and applicability. It should be clear whether the data provide fire model validation, the design and testing of a new detection system, or evaluation of a new performance-based design. In addition, brief summaries of any previous tests that have occurred prior to the current test series are also included.

Objectives

The objective section outlines the need for the specific test series. The objectives of testing for performance-based designs are usually to provide the designer with material data or physical fire properties when the system is exposed to a specific fire size and type. The test data can be used in a performance-based design calculation, as a fire model input to evaluate the findings of a design, or for model validation.

Area of Concern Description and Design

Once the objectives of the testing are established, specifics on the building or system configuration to be tested are developed. The test setup may include the layout and characteristics of the area of concern, or as little as a specific item to be tested, such as a couch or piece of material from a mattress. The specifics of the area of concern can then be compared to the test setup, which, due to testing constraints, may not be an exact replica of the actual intended end use configuration.

Fire Protection Systems and Equipment

The fire protection systems that will be used in the test are described along with the current level of protection they provide. This establishes a benchmark of performance that must be met. In addition to the fire protection systems, other applicable systems, such as ventilation, are described.

This establishes the conditions that are present in the structure or space of concern. In some cases, the facility may not exist yet, as may be the case in designing the systems for a new ship. It may be difficult to obtain information on specific systems in such a case. A degree of engineering assumptions must be made along with a proper analysis on the sensitivity of the unknown variables.

Test Setup

Once the test area has been described, the test apparatus or test mock-up can be selected or designed to suit the needs of the testing. It is desirable to reduce the number of variables by closely matching the actual end use layout. However, in many cases this is not possible due to size constraints or the systems that are available.

In many cases, scaled mock-ups of the test compartments are created with the assistance of fire models. Along with the general shape and size of the test compartment or apparatus, the fire protection systems and fire sources that will be used need to be described as part of the test setup.

Fire Scenarios, Fuel Packages, and Configuration

Applicable fire scenarios are described with specific detail (such as fuel loads, dimensions, and ignition sources) to provide information relative to the design fires used in the performance-based analysis.

Measures of Performance

The measures of performance are used to establish the desired findings and reflect the objectives of the test series. They equate to the data needed to complete the performance-based analysis. They could include time to equipment failure, tenability criteria, cook-off temperatures, fire spread, smoke spread, or any number of measures.

Instrumentation

The instrumentation used to obtain the data and classify the outcome of the measures of performance is listed in this section. Typical instrumentation often includes thermocouples (TCs) that are used to measure temperatures. TCs can be used in an array (distributed horizontally or vertically), or a single TC can be used to measure temperatures in specific locations (e.g., within the test assembly or to monitor ambient conditions.)

Measurements of the optical density, gas concentration, heat flux (calorimeters/radiometers), vent flow (bidirectional probes), fuel mass loss and flow rate, and suppression system flow rates could be recorded. Video of the events is typically recorded. Wind speed and direction are also recorded during full-scale testing if wind could affect the outcome. The ambient temperature, humidity, or barometric pressure can also impact combustion rates or smoke movement and may be measured and recorded.

Data are generated by the installed instrumentation. The instrumentation is used to measure any number of conditions within and affecting the test space. Every variable that may affect the outcome of a given test must be considered. If a parameter is deemed inconsequential to the outcome or considered a known fixed value, it may not have to be measured.

Variables that are deemed necessary to measure require instrumentation. The instrumentation needed is dictated by the objectives and measures of performance. Usually, additional instrumentation is added to generate more complete data to better understand the system or assembly performance during a fire exposure.

In almost all cases, the necessary instrumentation can be found commercially. The upper and lower bounds of the instrumentation must be known to ensure that the equipment will provide the range of values necessary to complete the measurement. In addition, it is desirable to know that the equipment will function in potentially harsh fire environments or use precautions to provide survivability of the instrumentation.

The instrumentation will generally provide a voltage (or current, frequency, etc.) output that corresponds to a physical reading. This voltage output is fed into a data acquisition (DAQ) system that processes and records the reading. Considerable time can be spent setting up the DAQ to ensure that the instrumentation is checked and provides accurate readings. Post-processing is often necessary to convert the voltage outputs to measurements of physical phenomena.

Test Procedure and Test Matrix

This section outlines the procedures used to replicate the desired design scenarios and establish safety procedures. The test matrix identifies the variables to be changed between each test. The test matrix may change considerably as test results are collected and analyzed during the test series. Deviations from the original test plan should be documented and justified.

Test Results

The test results section is where the test report separates from the test plan. This section presents the results relevant to the measures of performance. Additional data may have been taken but are left out of this section and placed into an appendix for brevity purposes. This section can be combined with the analysis section, but for the most part it only presents the relevant data. The majority of the section is reserved for data tables and graphs.

Test Data Analysis

Collected data must be analyzed. This section provides analysis of the data to address the measures of performance. This may address the abilities of a suppression system, ventilation scheme, fire fighting tactics, spread of fire or smoke, or provide data on ordnance cooling. This may be as simple as graphing and determining peak temperatures or visibility thresholds.

It can also span to the other side of the spectrum and become quite complex and time-consuming, with the data being used to implement multiple runs in a computer model or applied as input into complex mathematical calculations, or even used to develop and validate new calculations. This may involve curve fitting, statistical analysis, uncertainty analysis, or other complex analysis techniques.

The level of analysis depends on the desired results needed for use in the performance-based design. Once the data have been analyzed, they are available to the engineer as a tool that can be referenced to complete a performance-based design. From this analysis, conclusions and recommendations can be made.

Test Report Contents

Once testing is complete, a test report is created to document every detail of the test series. The report begins in a similar fashion to the test plan, with an introduction, objectives, geometry description, fire fighting systems, and equipment sections.

These sections will change little between the test plan and test report. The test setup description may be adjusted to reflect the actual implementation of the equipment and any changes made to the test setup. Occasionally, testing is an iterative process where the test conditions are modified based on observations from each successive test case. This includes the fire fighting systems and equipment, fire scenarios, fuel packages, and configuration. The instrumentation, test procedure, and test matrix will change when applicable.

CONCLUSION

The traditional approach to standard fire testing establishes product performance from a given fire test. The ability of the product to perform within the safety bounds of a standard is deemed to maintain satisfactory levels of fire safety. This practice has been in place for so long that it provides the following:

- Specific criteria potentially give a level of comfort to the person applying the criteria.
- Manufacturers know what is required to comply with the specifications.
- The requirements are simple to apply.

Although not stated, the use of the prescribed codes is assumed to ensure an adequate level of safety. Manufacturers and fire protection engineers often require the flexibility to choose how overall safety requirements are to be met as new materials, components, and products are developed. The increasing implementation of performance-based design also requires greater flexibility and more specific test data.

It is the responsibility of the fire protection engineer who is developing the alternative approach to state explicitly the approach used to substantiate a design having an acceptable level of safety. One way to generate valid assumptions is to use a performance-based approach based on test methods that provide data in engineering units suitable for use in fire safety engineering calculations.

These data have no significance until they are correctly implemented into a fire hazard analysis or performance-based design. With performance-based designs and the expanding amount of test data, it is important to understand the implications of the test method used and its relationship to the design, and not rely on a generic reference to a standard test method with limited applicability.

REFERENCES

ASTM, *Standard Test Method for Surface Burning Characteristics of Building Materials*, ASTM E 84, ASTM International, West Conshohocken, PA, 2013a.

ASTM, *Standard Test Methods for Fire Tests of Building Construction and Materials*, ASTM E 119, ASTM International, West Conshohocken, PA, 2012a.

ASTM, *Standard Guide for Room Fire Experiments*, ASTM E 603, ASTM International, West Conshohocken, PA, 2013b.

ASTM, *Standard Guide for Measurement of Gases Present or Generated during Fires*, ASTM E 800, ASTM International, West Conshohocken, PA, 2007.

ASTM, *Standard Test Method for Determining Material Ignition and Flame Spread Properties*, ASTM E 1321, ASTM International, West Conshohocken, PA, 2013c.

ASTM, *Standard Test Method for Heat and Visible Smoke Release Rates for Materials and Products Using an Oxygen Consumption Calorimeter*, ASTM E 1354, ASTM International, West Conshohocken, PA, 2013d.

ASTM, *Standard Test Methods for Determining Effects of Large Hydrocarbon Pool Fires on Structural Members and Assemblies*, ASTM E 1529, ASTM International, West Conshohocken, PA, 2013e.

ASTM, *Standard Test Method for Determination of Fire and Thermal Parameters of Materials, Products, and Systems Using an Intermediate Scale Calorimeter (ICAL)*, ASTM E 1623, ASTM International, West Conshohocken, PA, 2011.

ASTM, *Standard Test Method for Measuring Smoke Toxicity for Use in Fire Hazard Analysis*, ASTM E 1678ASTM International, West Conshohocken, PA, 2010.

ASTM, *Standard Test Methods for Measurement of Synthetic Polymer Material Flammability Using a Fire Propagation Apparatus (FPA)*, ASTM E 2058, ASTM International, West Conshohocken, PA, 2013f.

ASTM, *Standard Practice for Full-Scale Oxygen Consumption Calorimetry Fire Tests*, ASTM E 2067, ASTM International, West Conshohocken, PA, 2012b.

ASTM, *Standard Test Method for Room Fire Test of Wall and Ceiling Materials and Assemblies*, ASTM E 2257, ASTM International, West Conshohocken, PA, 2013g.

Averill, J., et al., *Occupant Behavior, Egress, and Emergency Communication, Federal Building and Fire Safety Investigation of the World Trade Center Disaster*, NIST NCSTAR 1-7, National Institute of Standards and Technology, Gaithersburg, MD, 2005.

Babrauskas, V., The Cone Calorimeter, in *SFPE Handbook of Fire Protection Engineering*, National Fire Protection Association, Quincy, MA, 2008.

Belles, D.W., Fisher, F.L., and Williamson, R.B., Evaluating the Flammability Characteristics of Textile Wall Covering Materials, in *International Conference of Building Officials (ICBO) Building Standards*, July-August 1987, vol. LVI, no. 4, pp. 8–14, 52.

Chapter 11

Performance-Based Design Documentation and Management

INTRODUCTION

The *SFPE Engineering Guide to Performance-Based Fire Protection* (NFPA, 2007) provides descriptions of the types of documentation that should be prepared by the design team. This includes the documentation associated with the fire protection engineering design brief, a performance-based design (PBD) report, specifications and drawings, and operations and maintenance manuals.

Fire Protection Engineering Design Brief

The development of goals, objectives, performance criteria, design fire scenarios, trial designs, and evaluation methods constitutes the qualitative portion of the design. Agreement of all stakeholders should be attained prior to proceeding to the quantitative analysis. A fire protection engineering design brief is suggested by the *SFPE Engineering Guide to Performance-Based Fire Protection* (NFPA, 2007) as a mechanism for achieving this agreement.

Evaluating and formally documenting performance-based designs can require extensive effort, and if fundamental aspects of the design change after detailed evaluation, significant rework may be required. For example, if property protection goals are later identified for a design completed and evaluated based on achieving life safety goals, then expended efforts may have been wasted. Similarly, if project stakeholders insist on certain types of design strategies, then these should be identified before other types of design strategies are developed and evaluated. The purpose of the fire protection engineering design brief is to facilitate agreement on the qualitative portions of the design prior to conducting detailed engineering analysis.

The contents of the fire protection engineering design brief will typically include the project scope, goals, objectives and performance criteria, design fire scenarios, and trial design strategies proposed for consideration. The form of the fire protection engineering design brief is intended to be flexible, based on the needs of the project and the relationship of the engineer performing the design to other stakeholders. In some cases, a verbal agreement may be sufficient. In other cases, formal documentation, such as minutes of a meeting or a document that is submitted for formal review and approval, may be prudent.

Once the design team and stakeholders have agreed on the approach that is proposed for the performance-based design, the detailed analysis work begins. This includes quantification of the design fire scenarios, evaluation of trial designs, and development of project documentation.

Design Documentation

Following completion of the evaluation and selection of the final design, thorough documentation of the design process should be prepared. This documentation serves three primary purposes: (1) to present the design and underlying analysis such that it can be reviewed and understood by project stakeholders, such as regulatory officials, (2) to communicate the design to the tradespeople who will implement it, and (3) to serve as a record of the design in the event that it is modified in the future or if forensic analysis is required following a fire.

The guide suggests that a detailed performance-based design report should be prepared that describes the quantitative portions of the design and evaluation. Every model or calculation method that was used should be identified in the documentation, including the basis for selection of the model or calculation method. Similarly, any input data for the model or calculation method should be identified, including the source of the input data and the rationale of why the data are appropriate for the situation being modeled.

All fire protection analyses have some uncertainty associated with them. The design should include methods of compensating for this uncertainty, and how this was accomplished should be documented.

As with prescriptive designs, performance-based designs use specifications and drawings to communicate to tradespeople how to implement the design. However, master specifications may not be applicable to performance-based designs without significant editing. Similarly, any features of a design that differ from typical prescriptive designs should be clearly identified on drawings.

One feature of documentation of performance-based designs that differs significantly from prescriptive-based designs is the operations and maintenance (O&M) manual. The O&M manual is described later in this chapter in more detail.

PERFORMANCE-BASED DESIGN MANAGEMENT

Fire protection systems in buildings constructed using performance-based methods are based upon building characteristics, occupant characteristics, and fire characteristics. When a prescriptive code is applicable, the design engineer must prepare documentation stating what deviations have been made, the rationale behind these alterations, permitted occupancy and load characteristics, how to maintain the fire protection systems, and how to inspect them for operation compliance. With the parameters established in the performance-based design, these documents exist for the benefit of the owner, occupant, and enforcement officials to create an environment that is as safe as prescriptive code requires and to permit future approval of occupancy as well as flexibility in use.

As a fire protection professional, it is important to understand the facility design and use progression as it pertains to the lifetime of a performance-based structure. This process includes fire protection design development, construction, certificate of occupancy, inspection, and possible changes or renovations to the facility.

PBD Building Lifetime

Figure 11.1 provides a depiction of the building design, construction, and use process or lifetime. The life of a structure begins at the feasibility study and design. Once the design is complete, the design must be accepted before it can be implemented. Then, construction begins and is completed. The authority having jurisdiction (AHJ) then reviews the construction and system commissioning. Once satisfied with the construction, the AHJ issues a certificate of occupancy, allowing the owner to occupy the structure. Then the owner is responsible for maintaining the required building features. In addition, a code official might visit and inspect the structure to verify that it is in compliance with the required criteria. Due to funding issues, many local governments do not have the personnel to perform this activity for buildings in their jurisdiction.

After some time, the use of the building may change. For example, the storage placed in a warehouse, chemicals used in a lab, or displays in an art gallery may change. Another example is that an owner may sell the building, and the occupancy may change from warehouse to a retail store, from office spaces to storage, or from business to assembly. As a result of these changes, different criteria may potentially become applicable.

When utilizing PBD, the building design, construction, and use process potentially changes in several areas when compared to a prescriptive-based design. Areas in which changes could occur include the following:

- The acceptance/approval process, in both design reviews and acceptance testing.
- The tracking of the PBD once the certificate of occupancy is accepted, including inspections by the AHJ once it is occupied.
- Operations and maintenance of existing building features by the owner.

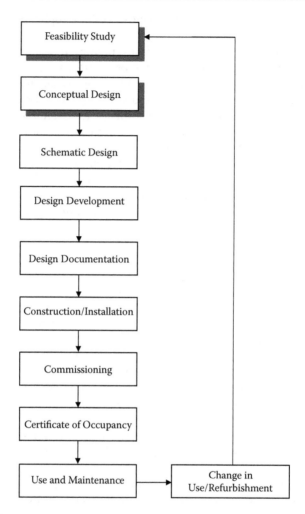

Figure 11.1 Basic building design and construction process. (From NFPA, *SFPE Engineering Guide to Performance-Based Fire Protection*, National Fire Protection Association, Quincy, MA, 2007. Used with permission of Society of Fire Protection Engineers, copyright © 2007.)

- When a change in use or occupancy is proposed, the owner and AHJ will require an analysis of the effects on the structure. In a prescriptive design, change in use requires a reanalysis relative to code compliance. In a PBD, a reanalysis of the structure is required based on the performance of existing systems for the new occupancy or usage.

ACCEPTANCE/APPROVAL PROCESS

The acceptance/approval process for a performance-based design varies from a prescriptive approach in that it may require different resources for the enforcement officials and the owner. These resources include the following:

- Administrative capabilities. The ability to track and retain the documentation becomes an administrative issue. The documentation should include agreements reached and how they are incorporated in the design.
- Personnel. Performance-based designs typically require more time to develop and review than prescriptive designs. Many hours and meetings are typically associated with understanding the design and its concepts.
- Technical capabilities. A performance-based design typically requires greater technical capabilities to understand the concepts associated with the design. Many enforcement officials do not have a fire protection engineer on staff.

Performance-based designs represent additional work for the enforcement officials and the building owner, and they potentially require more understanding of fire safety concepts than applying prescriptive requirements.

Administrative Capabilities

Administration capabilities include coordinating meetings, deliverables, reviews, and correspondence. For larger projects, a performance-based design may result in more than 30 meetings and lots of paperwork, e.g., meeting minutes, reports, specifications, and drawings.

Administrative issues can arise for the enforcement official and the owner. To be constructed, maintained, and occupied appropriately, documentation must be maintained by either the owner or the enforcement official throughout the life of the structure. This documentation includes (NFPA, 2007):

- Design reports
- Specifications and drawings
- Deed restrictions
- Identification of areas requiring special inspection and testing
- Operations and maintenance manuals
- Special inspections and testing reports
- Verification of compliance

Methods to maintain this documentation include for the enforcement official to establish files, and possibly deed restrictions, for each building, or for the owner to maintain all documentation.

Documentation should usually be kept in the building. When stored in the building, it is always accessible to inspectors or others who may need it. The documentation is also more likely to transfer to the new owners if the building is sold. However, duplicates should also be stored safely in an off-site location. This precaution is instituted so that if there is a fire or other mishap at the building, the documentation won't be lost.

Designers should also maintain the records associated with the performance-based design. The basis for a design may rest in the meeting minutes associated with the agreement of the design fire scenarios. The designer's liability may directly be affected by what the documentation states and who agreed to it.

Personnel and Technical Capabilities

As the design is developed, both the enforcement official and the owner should develop an understanding of several elements. Both the enforcement official and the owner should develop an understanding of the technical elements of the proposed approach. Additionally, they should also understand the implications of the limitations on the use of a building associated with the design.

Enforcement Official

To review a design, an enforcement official should have technical capabilities to understand the limitations associated with the design, review reports, review model(s) used and their input and output, and understand the results. Many enforcement officials will not have a fire protection engineer on staff, which can make it difficult for them to review a performance-based design.

A design in which sprinklers are omitted on the premise that occupants can egress before hazardous conditions develop may include criteria about fire growth, walking speed, tenability limits, combustion products, smoke production, heat flux, and fuel load. Each of these criteria should be verified or checked against literature values, which can be a time-consuming process.

Once fire models are introduced, additional challenges can arise. Computer programs are often flexible enough to compute solutions to problems outside the limitations of the model. The enforcement official should be familiar with the programs used by the designer and whether they have been verified and validated for a particular scenario or application. For example, model assumptions may be conservative for one situation, but they may be nonconservative for another. The code official should have a clear understanding of the assumptions utilized by models used in designs.

In addition to maintaining a grasp on the theory of the solution, the enforcement official should also inspect the design plans for code compliance. Fundamentally, performance-based codes are meant to provide an equivalent or greater level of safety than prescriptive codes. If the design is developed as an alternative to compliance with prescriptive codes, the code official should be able to understand the level of performance achieved by prescriptive codes. That is, the official should have an understanding of the code, rather than applying a checklist-type approach for prescriptive design.

The SFPE has documented a method for both code officials (ICC, 2004) and designers (NFPA, 2007) to use the approval process. For a code official to accept a proposed design, the enforcement official should understand how the solution works. Understanding the solution requires that the enforcement official is provided with documentation that shows the level of safety. As a practical matter, communication difficulties may exist between engineers and officials due to differences in technical backgrounds and professional responsibilities.

One of the first steps for the enforcement official is to determine if he or she has the personnel and technical capabilities to review the analysis. With all the duties an enforcement official must perform, it is possible for the official to become overwhelmed by the amount or nature of work. Therefore, an enforcement official may hire third parties to inspect plans and facilities.

The enforcement official has the option of consulting a qualified third party to have the design proposal reviewed at the expense of the designer or owner when the proposal is beyond the technical expertise or workload capacity of the enforcement official. Third-party reviewers should have education experience that demonstrates that they are competent to perform the peer review (SFPE, 2009). The third party is charged with the duty of evaluating the design and should report findings back to the enforcement official.

Two varieties of reviewers available include *contract* and *peer* reviewers.

- Contract reviewers are chosen by the enforcement official; i.e., the enforcement official makes the decision as to which contract reviewer to hire without input from the building owner or other stakeholders. The enforcement official also has the option of selecting a contract reviewer similar to soliciting a government work contract, entailing a formal request for offers and applicant screening. The contract reviewer is expected to review the entire project, to confirm that all building codes and standards have been satisfied with respect to both prescriptive and performance requirements (ICC, 2004).
- Peer reviewers are "peers" to the designer that submitted the design proposal. Detailed guidelines are established by the SFPE to perform peer reviews (SFPE, 2009). Peer reviewers should be qualified to create a fire protection design proposal that accomplishes the same objectives and satisfies the same criteria under the same limitations as the

original firm. They should also be able to demonstrate the comparable expertise to the original designer through documentation of experience and education. Peers should be examined for conflict of interest and bias—both of which should be addressed. The owner can submit multiple peers to the enforcement official for consideration, thus ensuring that the interests of the owner are kept in mind (ICC, 2004).

- Peer reviewers should be able to evaluate any technical aspect of a design, including components, design brief, conceptual approaches, recommendations, application or interpretation of code requirements, or supporting analyses and calculations (SFPE, 2009). Reviews should verify that both the correct questions are being asked and the questions are being answered correctly.
- The report produced by the peer reviewer details much of the original design proposal. The *SFPE Guidelines for Peer Review in the Fire Protection Design Process* (2009) suggest the following items should be addressed in the report:
  - Applicable codes, standards, and guides
  - Design objectives
  - Assumptions made by designer (e.g., performance requirements, design fire scenarios, material properties used in correlations or models)
  - Technical approach used by the designer
  - Appropriateness of models and methods used to solve the design problem
  - Input data to the design problem and to the models and methods used
  - Appropriateness of recommendations or conclusions with respect to the results of design calculations
- Correctness of the execution of the design approach (e.g., no mathematical errors or errors in interpretation of input or output data)

Although contract and peer reviewers both serve to advise the enforcement official, they may do so in differing scopes or using different contract mechanisms. That is, the contract reviewer fulfills a broad role, and is typically under contract to the enforcement official, whereas the peer reviewer potentially reviews specific elements and may be paid by the designer. Peer review and contract review are methods to address personnel and technical limitations associated with the review of performance-based designs.

Designers should allot sufficient time for the enforcement official to review the plans in depth. In all practicality, the review time allowed is usually driven by the fast-track nature of design and construction projects. Timing is almost always squeezed as construction grows closer. That is why it is important to agree to and establish timing for submittals and reviews and explore methodologies to expedite reviews, e.g., paid overtime or hiring a peer reviewer.

Owner

The owner also needs to be informed of the practical limitations associated with design assumptions. Without having a fire protection background, the owner may agree to limitations because he or she sees it as a means to an end. The owner may agree to limit fuel loads to satisfy his or her design goals.

However, the practical limitations of the design should be fully explained to the owner. For example, if the owner realizes decreased revenues because he or she can't put a fiberglass boat in the atrium as advertising, the owner needs to know that. Or if an owner can't use the atrium floor for trade shows, that may be a concern for him or her. In addition, increased maintenance costs can offset savings.

CHANGES DURING CONSTRUCTION

Unexpected changes are a fact of life in the construction business. Changes in the field can be caused by misalignment of assemblies, carelessness, inconsistencies in drawings, or other factors. For instance, rated or listed assemblies require attention to detail in order to preserve their listing. In some cases, separation assemblies between occupancies or residential apartments may not be constructed appropriately.

In case of mistakes like these, the enforcement official may require analyses to demonstrate that the structure, as built, still satisfies performance requirements, or he or she may demand compensatory action, e.g., that portions of the structure be rebuilt. Based on changes during construction, additional engineering analysis may be needed to determine whether acceptable performance would still be expected.

Another type of change comes on the drawing board itself. Because many large structures are very complicated, the work may be partitioned and distributed to specialty firms or departments within larger companies. The individuals to which these jobs are delegated may have expertise in their own fields, but usually do not have expertise in every field. Therefore, it is possible for essential features to be inadvertently eliminated in the name of value engineering or cost reduction because other designers fail to see the need for a particular component.

For example, dampers to allow fans to provide 100% outside air supply may be inadvertently omitted. Outside air may require conditioning, and a mechanical engineer who did value engineering may have thought that these dampers were unnecessary. However, the fans are also used for smoke control, and the dampers may have allowed unconditioned air into a space when the smoke control system is in use. The resulting costs of installing the dampers may exceed any savings.

Changes resulting from construction variations or value engineering should be controlled to verify they do not affect the fire safety approach for the structure. Ideally, controls would be instituted to prevent such oversights. However, mistakes will probably be made and not discovered until the situation makes change difficult.

To address concerns, the enforcement official or engineer can require reconstruction, or require some sort of compensatory action. This compensatory action may entail something as simple as an extra layer of wallboard or as complicated as redesigning an as-yet uninstalled fire protection system to improve performance. Alternatively, installed systems may need to be upgraded to provide more flow, earlier detection, etc. All stakeholders should be consulted to determine the best course of action.

Engineers should not assume that others are keeping track of required fire safety systems. In most cases, the person managing the money has a significant contribution to the decisions that are made.

It should be noted that although value engineering strives to reduce costs, uninformed engineers can increase the cost of projects precisely by making value engineering mistakes. To minimize the occurrence of costly errors, oversight needs to be instituted. One possibility is to annotate drawings with critical requirements. Another is to use multiple reviews to create an overlap of expertise. In the case that work is spread among different firms (and therefore different management chains), copies of final proposals should be sent to every firm involved for their oversight.

CHANGE IN USE

For a system built upon a specific use of a space, post-construction changes in the use may require reevaluation, independent of whether the design is based on prescriptive- or performance-based criteria. More specifically, whatever limitations are placed on a design must be maintained.

For example, a prescriptive-based building designed as an office building can't be changed to a conference center without analysis of how the change in use affects code criteria. An office building is classified as a business occupancy. A conference center is an assembly occupancy with different construction, fire safety system, and egress criteria.

Similarly, a change in use of a structure can affect a performance-based design. However, with a performance-based design, even a minor change may have an impact on building performance. A change in use could involve different contents, different activities, or different construction features.

For example, the conference center in the example above will have more occupants, but depending on the number of doors, the egress time may remain unchanged. In addition, if occupants have a single focus, they may be able to react faster. The effect on fuel load will vary depending on the use. All these factors may need to be reviewed in a performance-based design.

The operations and maintenance (O&M) manual (described later in this chapter) should identify the limitations associated with the performance-based design. It should also identify allowable alterations. Alterations that fall outside of those initially determined to be allowable require an analysis to determine if the building still complies with the intent of the performance-based design. The structure may still be safe if the change went beyond the bounds of the O&M manual, but that must be determined by an analysis.

Costs of change in use may also be higher due to the additional engineering work required. For example, if one of the fuel load limitations is exceeded, more documentation may be required from the engineering team, and the potential exists for reanalysis by the original engineers and any other third parties.

OPERATIONS AND MAINTENANCE MANUAL

Many building features in a performance-based design will be identical to those in a prescriptive design. However, certain features will be unique. If changes to those features could impact building performance, those features should be identified in an operations and maintenance (O&M) manual.

The operations and maintenance manual communicates to facility managers the limitations of the design. These limitations stem from decisions made during the design process. For example, heat release rates used as input data place a limitation on the use of a space. Any furnishings placed within a space that could have higher heat release rates than used during fire modeling could result in more severe consequences than the model predicted.

The operations and maintenance manual should be written in a format that can be easily understood by people who are not fire safety professionals, since most building owners and facility managers will not have this type of background. Typical contents of an O&M manual are provided in Table 11.1 and summarized in the following sections.

Operational Requirements

Fire safety systems designed using performance-based methodologies can be different from prescriptive systems in terms of their operation, upkeep, and alteration. Compared with prescriptive systems, performance-based systems can place further limitations on how the building may be changed or used.

In some sense, performance-based designs are the sports cars of fire protection design and may require a higher level of detail in maintenance and capabilities than a typical prescriptive system to obtain higher performance. Therefore, the design team should inform the interested parties of

Table 11.1 Sample O&M Manual Contents

1. As-built drawings of building and systems
2. Information on installed systems
   2.1. Installing contractor
   2.2. Maintenance contract
   2.3. Product data sheets and specifications
   2.4. Maintenance manuals
3. Limitations on use of building
4. Testing and maintenance requirements
   4.1. Method of testing
   4.2. Frequency of testing
   4.3. Documentation of testing
5. Compensatory measures
   5.1. Identification of critical systems
   5.2. Identification of what actions must be taken for various conditions of system impairment
6. Control of combustible loading
   6.1. Description of combustible loading contemplated by design
   6.2. Process for quantifying changes in combustible loading
   6.3. Threshold values at which additional protection features/practices must be implemented
7. Allowable alterations
   7.1. Description of building uses and arrangements contemplated by design
   7.2. Description of allowable tenant alterations
      7.2.1. Maximum size of spaces
      7.2.2. Allowable finish materials
      7.2.3. Allowable occupancies
      7.2.4. Maintenance of exits
   7.3. Inspections by code official (or designated representative)
      7.3.1. Frequency
      7.3.2. Scope and procedures
      7.3.3. Inspection forms

Source: ICC, *The Code Official's Guide to Performance-Based Design Review*, International Code Council, Washington, DC, 2004. Used with permission of Society of Fire Protection Engineers, copyright © 2004.

the upkeep requirements in the form of an operations and maintenance (O&M) manual (NFPA, 2012).

In general, the O&M manual lists the requirements of the building operator to verify that the entire system and its subcomponents operate properly and in accordance with the design, thus giving the anticipated level of performance. This includes a maintenance schedule consisting of inspection, testing and maintenance frequency, procedures, and expected results. Special attention should be paid to particularly unusual components while the O&M manual is being written.

If a performance-based design is based on a maximum heat release rate, e.g., 2,000 kW, or total smoke production rate, the O&M manual should establish the criteria that the enforcer or owner can verify. For example, the

O&M manual may state that only one 4 ft (1.2 m) long California foam couch may be located in the lobby unless it is separated by more than 15 ft (4.6 m) from any other combustibles.

The O&M manual should also list activities that may not be performed within the building. This includes any activity that may invalidate the design basis or contradict the design assumptions.

The manual should also state the performance specifications considered by the designers. For instance, criteria such as "shut down computer room 8 s before gas discharge" or "detect flame before fire reaches 20 kW fire size" should be explicitly listed. It is useful to have this type of window into the mind of the engineer when changes to the building are planned. This way, there is less ambiguity for the enforcement official and design team to deal with at later stages. Furthermore, stated performance objectives may be useful for a forensic analysis after a fire.

In a similar vein, all design assumptions should be stated. Examples include "no natural gas-powered forklifts" and "no cigarette smoking." The presence of a clear list provides the owner and enforcement official with a reference for compliance verification. This can in turn aid in obtaining occupancy certification.

Another topic that should be addressed in the O&M manual is a description of permissible renovations. This section should describe the uses of the building and arrangements envisioned during the design process, which can be used as a guide for future alterations. Allowable changes by tenants should be described, if relevant. This may encompass the maximum size of spaces, the finish materials used, occupancies, and maintenance of exits. For an open-plan office structure, the O&M manual might limit the type of desks allowed or may constrain their arrangement within the space.

Facility maintenance staff and tenants should be informed about the requirements and restrictions placed upon a building by a performance-based design. The permanent nature of the O&M manual allows occupancy and use limitations to be recorded for future reference and inspection.

Staff and tenants should be informed about permissible activities, types of fuels, and fire load in terms people who are not fire protection specialists can understand. Staff in particular should be adequately trained to maintain the fire protection systems and to verify that the space remains in compliance with the design specifications.

Both staff and tenants should be educated about systems or restrictions. For example, employees in an airport traffic control tower with one exit should be educated about the hazard associated with storage in that exit. Steps may be taken to limit the impact of most fires in a single exit tower. However, the effects of a fire in the exit will almost certainly impact occupants.

Compensatory measures should be developed and included in the manual to deal with incapacitation of system components before failure occurs. The rationale behind this is to prevent the improvisation of ad hoc, ill-informed measures while a (possibly critical) component of the system is inoperable.

To accomplish this, the critical systems should be identified, such as alarm control panels and smoke detectors. Next, the duration of incapacitation should be considered, because practical differences exist in the handling of a 2 h impairment and a 30 h impairment. Finally, some scheme to compensate for the specific problem should be proposed and evaluated. Some possible compensatory measures include institution of a fire watch, keeping an inventory of spare parts, or establishing maintenance contracts with specified response times in order to limit downtime of a system.

In reality, the owner is seldom anxious to hire someone for the life of the building to verify the performance-based design assumptions are maintained. However, as an engineer, it is a good idea to submit a proposal to provide inspection services with any performance-based design to assist in verifying the recommended actions are maintained throughout the life of the building. Therefore, at least the concern is identified and potentially addressed.

Inspection, Testing, and Maintenance

The design documents and O&M manual should identify the inspection, testing, and maintenance requirements associated with maintaining a performance-based design. Many performance-based designs may have criteria for testing and maintenance of systems that are similar or identical to prescriptive designs. However, some designs may be predicated on more rigorously designed and tested systems to achieve the desired objectives. Inspection testing and maintenance can increase reliability in the system.

Before a certificate of occupancy is issued, but after construction has been completed, the enforcement official may inspect the fire protection systems. The official may verify compliance with the design basis, assumptions, and specifications for performance-based designs, in addition to compliance for prescriptive components of the system. The O&M manual is a useful tool for this purpose because it will contain information on the method of testing and acceptable results for each system. The O&M manual should also include specifications for the frequency of testing and sample forms to record results and verify that the tests were performed correctly.

The enforcement official may make initial inspections of sprinkler, alarm, and passive fire protection systems, which requires a wide range of expertise. The enforcement official may also be concerned about certificates of occupancy and allowable use of the building. Furthermore, the enforcement official may enforce stipulations in the O&M manual about usage, load, or maintaining design assumptions. This may require periodic visits to the site and familiarity with the design documentation.

The enforcement official may also play a role in periodic testing. The local official may test fire safety systems on a regular basis in all buildings in the jurisdiction. In a performance-based designed building, the enforcement official may choose to visit more often, giving the official the opportunity to verify maintenance of site features.

In reality, very few enforcement officials have the personnel or funding to allow frequent testing and visits. It may be desirable to use someone other than the local enforcement official to perform inspections. In this circumstance, the O&M manual can detail qualifications or a method of selecting third-party inspectors.

The results from all tests should be documented and retained with the building records. This allows degradations in performance to be noticed and repaired, or might give early warning of impending failure.

REFERENCES

ICC, *The Code Official's Guide to Performance-Based Design Review*, International Code Council, Washington, DC, 2004.

NFPA, *SFPE Engineering Guide to Performance-Based Fire Protection*, National Fire Protection Association, Quincy, MA, 2007.

NFPA, *Fire Code*, NFPA 1, National Fire Protection Association, Quincy, MA, 2012.

SFPE, *SFPE Guidelines for Peer Review in the Fire Protection Design Process*, Position Statement P-03-09, Society of Fire Protection Engineers, Bethesda, MD, 2009.

Chapter 12

Uncertainty

INTRODUCTION

The *SFPE Engineering Guide to Performance-Based Fire Protection* defines *uncertainty* as "the amount by which an observed or calculated value might differ from the true value" (NFPA, 2007). In engineering, there are two types of uncertainly: epistemic and aleatory.

Epistemic uncertainty is uncertainty due to lack of (complete) knowledge. For example, it may not be possible to calculate precisely what the temperature would be in a post-flashover fire due to approximations used in models and input values. It is possible to develop estimates; however, these estimates will not be exact, even if the input variables are well known. Epistemic uncertainty can be reduced by gaining additional information or knowledge.

Aleatory uncertainty is uncertainty due to random variation. For example, sprinklers that are manufactured may have slight variation in activation temperature and response time index (RTI). It might be known that these factors will vary by no more than a certain percentage, but it is impossible to reduce this uncertainty without changing the way the sprinklers are manufactured. Whether a coin toss will result in heads or tails is a form of aleatory uncertainty. No amount of additional research or information would reduce this uncertainty.

In fire protection engineering, uncertainty primarily arises from the following sources (NFPA, 2007):

- **Theory and model uncertainties.** Models or correlations are frequently used in fire protection engineering analysis. While these models and correlations provide results that make performance-based design possible, they may also introduce uncertainty into the design process. Calculation methods may be based upon an incomplete or less than perfect understanding of the underlying science. For example, an empirically derived correlation may be based upon data from experiments that were conducted under a certain range of conditions. Without additional analysis, it is unknown how well the correlation would perform outside of this range. Similarly, in some cases simplifications may be incorporated into a model or correlation based on an assumption that it will only be used within a limited set of conditions. The solution routines incorporated into National Institute of Standards and Technology's (NIST) Fire Dynamics Simulator are based on simplifications made possible by assuming low Mach number flows. While this is a reasonable assumption in most fires, it would not be reasonable for modeling explosions.

- **Data and model inputs.** The data used as input into models or correlations are subject to uncertainty. For example, it may not be possible to precisely determine some material properties. Consider flashpoint— for any given flammable or combustible liquid, use of an open-cup apparatus may yield a different measurement than use of a closed-cup apparatus. As input data become more complex, uncertainty can also become magnified. Consider burning rates used as input into a fire model. If the data used were gathered in a manner that is not similar to the scenario being modeled, increased uncertainty may result.

- **Calculation limitations.** In cases where the science underlying a model or correlation is well understood, simplifications may be incorporated to enable the model or correlation to be solved in a reasonable amount of time. For example, consider techniques for modeling room fires. Both zone models and field models are approaches for solving complex simultaneous equations of conservation of mass, momentum, and energy. The discretization technique employed in both types of models results in solutions that are approximate, but never exact.

- **Design fire scenario selection.** Design fires are generally used for the analysis of performance-based designs. Uncertainty may be introduced if the design fire scenarios selected do not represent the range of fires that might occur.

- **Uncertainty in human behaviors.** Human behavior is a highly stochastic phenomenon. There are a number of actions that people may take in response to a fire. Where human behavior is considered, uncertainty can be introduced due to its stochastic nature.

- **Uncertainties in risk perceptions, attitudes, and values.** This type of uncertainty pertains to the selection of objectives and performance criteria. In some cases, it may be difficult to accurately assess the level of safety that is desired by the stakeholders of a fire protection design. Determining the tolerable risk to life from fire in a building is a classical example.

METHODS OF TREATING UNCERTAINTY

Uncertainty can arise from a number or sources. However, there are a number of techniques that can be used to compensate for uncertainty in the design process.

Safety Factors

Safety factors are frequently used in performance-based designs to deal with uncertainty or provide a margin of safety between what would be considered acceptable and unacceptable outcomes. In most cases, these safety factors are developed on an ad hoc basis, since generally few fire protection engineering methodologies provide specific safety factors. For example, in egress analysis a safety factor of 2 seems to be frequently used (Fleming, 1999). However, since safety factors are critically important for the protection of public health, safety, and welfare, they deserve careful consideration.

In structural design, safety factors are applied to both increase the expected loads and decrease the resistance to the loads that a structure provides (Lucht, 1996). Multiplicative safety factors greater than 1 are used to overestimate loads, while multiplicative factors less than 1 are used to underestimate material properties. For example, in bridge building, safety factors are used to ensure that a bridge built of the weakest conceivable batch of steel would support the load associated with the heaviest possible truck crossing the bridge under the worst possible weather and traffic conditions (Petroski, 1994).

Ideally, if sufficient data are available, safety factors can be developed by comparing calculations to available data and determining how much the data vary from calculated values. Three of SFPE's engineering guides have developed safety factors in this manner (SFPE, 1999, 2000, 2002). Based on comparisons of predictions with available data, safety factors recommended when using the methods identified in the guides range from 1.5 to 3.7.

Safety factors are generally increased when public safety or welfare is at stake. For example, design structural loads are increased by importance factors, which are larger for buildings in which large numbers of people might congregate or buildings that perform vital public safety roles (Lucht, 1996). Similarly, the safety factor that is used for a cable that drives a dumbwaiter is smaller than that used for a freight elevator, which is in turn smaller than that used for a passenger elevator (Fleming, 1999).

Since there is generally more uncertainty in fire protection designs than in structural designs, safety factors are a critical component of performance-based fire protection design. However, safety factors should be established with care, since selecting too low a safety factor could result in a design that would not be safe under some conditions, while too large a safety factor would lead to a design that is overly expensive to build.

Figure 12.1 shows the basis for a safety factor of 2 when the Shokri and Beyler correlation is used to predict thermal radiation from a hydrocarbon pool fire. This graph compares data from measured heat fluxes versus predicted heat fluxes. The measurements were obtained from full-scale experiments, and the predictions were developed using the Shokri and Beyler correlation to model the experiments.

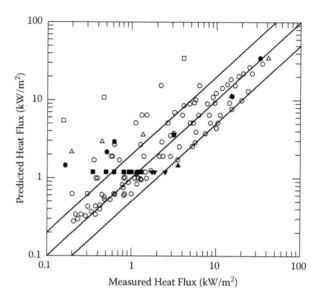

Figure 12.1 Comparison of measured and predicted heat fluxes. (From Beyler, C., *Fire Hazard Calculations for Large, Open Hydrocarbon Fires*, in *SFPE Handbook of Fire Protection Engineering*, 4th ed., National Fire Protection Association, Quincy, MA, 2008. Used with permission of Society of Fire Protection Engineers, copyright © 2008.)

The center diagonal line shows perfect agreement. If all data points fell on this line, then no factor of safety would be necessary. The other two diagonal lines are drawn at factors of 2 above and below the line of perfect agreement (i.e., multiplied by 2 and divided by 2). It can be seen in Figure 12.1 that most of the data fall within the two lines representing factors of 2, and the data points that fall outside of these lines represent cases where the correlation predicted a greater heat flux than was measured. For design, one generally wishes to overpredict a heat flux rather than underpredict it, as a larger heat flux would represent a greater potential for harm. Therefore, Beyler (2008) concluded that a factor of safety of 2 would be sufficient.

Analysis Margin

Analysis margin is the difference between the estimated value of a parameter and the value used in analysis. Larger margins correspond to larger confidence that the value will not be exceeded. However, larger margins also correspond to more expensive designs, so the analysis margin should be selected carefully.

For example, consider the mass of combustibles per unit area found in an office. Field surveys and literature searches would yield a range of values.

One could calculate a mean and standard deviation for this range. The mean would be the best estimate. The best estimate value is a value that would be expected to be exceeded for half of the cases, while in the other half of the cases the actual value would be less.

In deterministic analyses, best estimate values are generally undesirable. However, best estimate values are frequently used in risk-based analyses. For deterministic analyses, bounding values are frequently used.

A bounding value is a value that is never expected to be exceeded. A bounding value might be the highest possible or the lowest possible value for a parameter, depending upon the type of analysis. Consider estimation of post-flashover fire temperatures in a compartment. A bounding value for the mass of combustibles would generally be the highest possible value. However, a bounding value for the thermal conductivity of the lining materials would be the lowest possible value, since lower thermal conductivities would equate to higher compartment temperatures.

Sufficiently bounding is an approach that could also be used in deterministic analysis. If a calculation requires multiple parameters as input, and bounding values are used for each parameter, then the end result might be overly conservative. For example, if a calculation requires three parameters as input, and 95% confidence values are used for each value, then the end result would represent a 99.9875% confidence value (assuming that the values for each input parameter are independent of each other).

Sufficiently bounding is an approach where all but one input parameter is set to best estimate values, and the remaining parameter is set at a 95% confidence or bounding value. Sufficiently bounding could also be used for consequence estimation in risk matrix analyses.

Importance Analysis

For calculations that require multiple input variables, each input value might not contribute equally to the uncertainty in the calculation. For example, consider Alpert's correlation for temperature rise above a fire plume (Alpert, 2008):

$$\Delta T = 16.9 \frac{\dot{Q}^{2/3}}{H^{5/3}}$$

where ΔT is temperature rise above ambient (°C), $\dot{Q}$ is heat release rate (kW), and H is height above base of fire (m).

When performing a calculation using this correlation, both H and $\dot{Q}$ might be subject to uncertainty, with $\dot{Q}$ most likely subject to a greater uncertainly than H. However, since $\dot{Q}$ is raised to the 2/3 power, any uncertainty in the heat release rate is reduced in its effect on the result. Since the height above base of fire is raised to the 5/3 power, its uncertainty on the result is magnified.

Importance analysis rank orders the effect of uncertainty in each variable on the overall result. In the case of Alpert's correlation for temperature rise in a fire plume, the rank ordering would be (1) height above base of fire and (2) heat release rate.

Sensitivity Analysis

Sensitivity analysis is a technique that can be used to determine how changes in input variables affect the calculated result. A sensitivity analysis is conducted by holding all but one input variable constant while varying a single parameter and determining the effect.

The sensitivity of input variables is typically reported as a decimal, which is calculated as follows:

$$\frac{\text{Percent change in output}}{\text{Percentage change in input}}$$

When conducting a sensitivity analysis, a base case is selected (which is usually the case of interest) and each parameter is individually varied by a percentage (e.g., 10%, 50%, etc.). If the calculated sensitivity is greater than 1, then the output is more sensitive to that parameter. If the calculated sensitive is less than 1, then the output is less sensitive to that parameter.

Sensitivity analysis is useful because it provides insight into the parameters that have the potential to contribute the most to the uncertainty in the calculated result. Careful attention should be paid to the input parameters to which the calculated result is most sensitive. If the calculated result is more sensitive to the parameter, then the analyst may need to use either a more accurate value or a bounding value for that parameter. Conversely, if the calculated result is less sensitive to the parameter, then it is not as important to precisely determine the value.

For continuous functions, it is possible to conduct a sensitivity analysis by use of differential calculus. Consider a function $F(x_1, x_2, \ldots, x_n)$. If a base case is selected, and the derivative of the function is taken with respect to each variable (e.g., dF/dx_n), then graphs can be prepared of the sensitivity of each variable by plotting the derivative over the ranges of input variables of interest, e.g., dF/dx_n versus x_n. These graphs would show how sensitivity of output would be affected by each variable.

Switchover Analysis

In a switchover analysis, the values of one or more input parameters are varied to determine which values would cause the calculated result to change from a value that would be considered acceptable (i.e., meet performance criteria) to a value that would be considered unacceptable (i.e., does not meet performance criteria).

If all foreseeable combinations of input values would not cause switch-over, then the analyst can be confident in his or her conclusions. However, if some combinations of values could cause switchover, then the analyst should ensure that it would not be possible for a combination of factors that could cause switchover to occur.

Classical Uncertainty Analysis

Classical uncertainty analysis (NFPA, 2007) is possible in situations where there is a large amount of measurement or test data available for a parameter. Uncertainty in measurements could be comprised of two components: random error, or variation, and systematic error, or bias. Random errors are fluctuations that arise due to variation in manufacturing or limitations of measurement devices. Systematic errors are consistently in the same direction. Systematic errors may arise from problems with an experiment or in a manufacturing run.

Random error can be calculated using the following equation:

$$S = \sqrt{\sum_{i=1}^{n} \frac{\left(X_i - \bar{X}\right)^2}{n-1}}$$

where S is random error, and $\bar{X}$ is data mean.

Random error is also known as the sample standard deviation.

Systematic error can be difficult to detect, except when replicate tests are conducted. Unless it is specifically reported for an experiment, it is typically assumed to be zero.

The use of confidence intervals is an example of a classical uncertainty analysis. A confidence interval represents the likelihood that a parameter will be within a given range. For example, it might be reported that a parameter is within the range with 95% confidence. As a confidence interval grows in precision, e.g., 99%, or if the random error increases, the range gets larger. A value of 95% confidence is frequently used in engineering applications.

UNCERTAINTY IN THE PERFORMANCE-BASED DESIGN PROCESS

At present, no single, widely accepted methodology exists for dealing with uncertainty in the design process. Notarianni (2008) provides a recommended procedure for applying uncertainty analysis in the context of the performance-based design process described in the *SFPE Engineering Guide to Performance-Based Fire Protection*. However, Notarianni's procedure requires that the performance-based design be conducted on a probabilistic, or risk-based, process. Risk-based analyses are infrequently used in building designs.

The *SFPE Engineering Guide to Fire Risk Assessment* (2005) provides a methodology for addressing uncertainty in fire risk assessments. However, like Notarianni's method, it is based on the use of a risk assessment.

Lundin (1999) developed a method for compensating for uncertainty in predictive methods. Lundin's method entails two steps:

- Determine the model error and uncertainty based on comparisons of model predictions with experimental data.
- Create a method of adjusting model output such that adjusted output reflects model uncertainty.

When conducting an uncertainty analysis, the important task is to verify that for reasonably foreseeable variations of input variables and inaccuracies in predictive models, predicted outcomes could not change from an acceptable outcome to an unacceptable outcome. Generally, some combination of the tools described above is used, e.g., sensitivity or importance analysis to determine the parameters that most heavily dominate uncertainty, switchover analysis to determine if possible variations could result in an unacceptable outcome, and safety factors to provide a higher degree of confidence in the overall result.

REFERENCES

Alpert, R., Ceiling Jet Flows, in *SFPE Handbook of Fire Protection Engineering*, 4th ed., National Fire Protection Association, Quincy, MA, 2008.

Beyler, C., Fire Hazard Calculations for Large, Open Hydrocarbon Fires, in *SFPE Handbook of Fire Protection Engineering*, 4th ed., National Fire Protection Association, Quincy, MA, 2008.

Fleming, J., Safety Factor(s) and Deterministic Analysis, presented at Final Proceedings—SFPE Symposium on Risk, Uncertainty, and Reliability in Fire Protection Engineering, Society of Fire Protection Engineers, Bethesda, MD, 1999.

Lucht, D., Public Policy and Performance-Based Engineering, presented at Proceedings—1996 International Conference on Performance-Based Codes and Fire Safety Design Methods, Society of Fire Protection Engineers, Bethesda, MD, 1996.

Lundin, J., *Model Uncertainty in Fire Safety Engineering*, Report 1020, Lund University, Sweden, 1999.

NFPA, *SFPE Engineering Guide to Performance-Based Fire Protection*, National Fire Protection Association, Quincy, MA, 2007.

Notarianni, K., Uncertainty, in *SFPE Handbook of Fire Protection Engineering*, 4th ed., National Fire Protection Association, Quincy, MA, 2008.

Petroski, H., *To Engineer Is Human—The Role of Failure in Successful Design*, Barnes and Nobel Books, New York, 1994; referenced in Fleming, J., Safety Factor(s) and Deterministic Analysis, presented at Final Proceedings—SFPE Symposium on Risk, Uncertainty, and Reliability in Fire Protection Engineering, Society of Fire Protection Engineers, Bethesda, MD, 1999.

SFPE, *Engineering Guide: Assessing Flame Radiation to External Targets from Pool Fires*, Society of Fire Protection Engineers, Bethesda, MD, 1999.

SFPE, *Engineering Guide: Predicting 1st- and 2nd-Degree Skin Burns from Thermal Radiation*, Society of Fire Protection Engineers, Bethesda, MD, 2000.

SFPE, *Engineering Guide: Evaluation of the Computer Fire Model DETACT-QS*, Society of Fire Protection Engineers, Bethesda, MD, 2002.

SFPE, *Engineering Guide: Fire Risk Assessment*, Society of Fire Protection Engineers, Bethesda, MD, 2005.

Index

Milton Keynes UK
Ingram Content Group UK Ltd.
UKHW040059071024
449327UK00019B/660